Brandis Dietrich

Brandis Memo Forest Laws

Brandis Dietrich

Brandis Memo Forest Laws

ISBN/EAN: 9783337232009

Printed in Europe, USA, Canada, Australia, Japan

Cover: Foto ©berggeist007 / pixelio.de

More available books at **www.hansebooks.com**

MEMORANDUM

ON THE

FOREST LEGISLATION

PROPOSED FOR

BRITISH INDIA,

OTHER THAN THE PRESIDENCIES OF MADRAS AND BOMBAY,

BY

D. BRANDIS,

INSPECTOR-GENERAL OF FORESTS TO THE GOVERNMENT OF INDIA.

Dated 2nd August 1875.

INTRODUCTION.

Objects of Forest Legislation.—Forest legislation at present in India has three distinct objects :—

1st.—To facilitate the demarcation, protection, and good management of those public forests which are commonly known as reserves, Government, or State forests. A large proportion of these have already been demarcated, that is, separated by defined boundaries from other public or private lands : in some cases demarcation is in progress, and in others it has not been commenced.

2nd.—To secure a certain amount of protection of forest growth on lands which are either the property of Government or on which Government has certain rights, but which have not been formally demarcated. These lands are commonly known as open, unreserved, or district forests. To this class belong all lands in which Government has not relinquished its forest rights, whether they form part of the Government waste lands or are in the occupancy of individuals or communities.

3rd.—To authorise the levy of certain rates of duty on timber and other forest produce, and the control generally of timber and other forest produce in transit, the aim of this being the protection of the Government forests as well as the protection of the interests of persons engaged in the timber trade.

Briefly it may be said that the object is to secure the protection and control of demarcated public forests, a certain protection of undemarcated forests, and the control of timber and forest produce in transit.

2. *Review of forests to which the Forest Law will apply.*—Before discussing in detail these main objects of forest legislation, it will be necessary briefly to review the present state of the forests in the different provinces for the protection of which legislation is required. These remarks will be limited to the provinces and districts to which the new forest law will, in the first instance

apply. They will not include the forests of Mysore and Berar (though they are under British forest officers), as they have forest rules of their own. Nor will they include any of the leased forests in the Himálaya (Chamba, Bussahir, Native Garhwál) or the forest lands in the British districts of Ajmere and Hazára, the protection and management of which is provided for by special forest regulations.

At the outset, I beg to draw attention to the following passage of a memorandum on forest legislation which I submitted to the Government of India in September 1864: "It is not intended to enforce the prohibitions of the forest rules throughout the whole of those forest tracts where Government has not ceded its right in the growing trees, brushwood, and timber. But it is intended to demarcate certain tracts as reserved forests which are to be strictly conserved and where the rules are to be enforced. Wherever any existing rights will be interfered with by the establishment of these reserved forests, it is intended to give compensation, either in land or otherwise, to the parties concerned. It is evident that, under these circumstances, the final demarcation of the forests intended to be reserved will require considerable time, and in many parts these forest tracts have been so impoverished by constant mutilation of the trees, cutting, and burning, that, before being made fit to yield timber and other forest produce for the requirements of the country, they will have to be left in perfect rest for a series of years, and in some instances planting operations on a large scale will have to be undertaken. Until the demarcation of reserved forests is completed and until they are in a fit state to be worked, a certain supervision must be exercised over the great mass of forests in which Government has any rights; and in provinces where forests are scarce and poor the whole of the forests may have to be kept under continuous control for generations to come." And in October 1868, when submitting a revised draft Forest Bill in the place of the Act of 1865, I made the following statement: "In several provinces a commencement has been made to separate from the large extent of waste lands at the disposal of Government certain of the more valuable forest tracts and to place them under the exclusive control of the Forest Department. These tracts have been designated as reserved or State forests, and generally they have been, or will be, limited and demarcated by boundary marks. The other forest lands at the disposal of Government which were not included in the State forests have in those provinces been called unreserved or district forests. At the same time it was not everywhere deemed expedient to give up and abandon the control of Government over the remainder of the waste and forest which had not been included within the reserved or State forests. For, in the present state of matters, it is not possible to know whether the State forests already selected will be found fully sufficient for the requirements of the country and of export trade, or whether their area may not have to be increased hereafter. Nor is it possible to foresee whether the requirements of the agricultural population may not in future render desirable the formation of village forests to be managed under the general control of Government with the view of guarding against improvident working and devastation. In fact, it is not improbable that, as the development of the country progresses, there will be in India the same great classes of forest property which are found on the Continent of Europe, viz.:—

I.—State forests.

II.—Forests of villages and other communities and public corporations or institutions, such as churches, schools and hospitals.

III.—Forests of private proprietors."

"If this state of things were now in existence, legislation would be comparatively easy. The Indian Forest Law would, like the *Code Forestier* of France or other Continental Forest Laws, define the degree of State supervision which in the public interest should be exercised over the management of these different classes of forests, and would prescribe those measures for the protection and management of the State forests and forests of communities and public bodies that require the authority of special legislation. At present however matters are in a transition state and the separation of these different classes of forest has not yet been accomplished."

In 1864 the demarcation of the Government forests had hardly been commenced, and now the area demarcated as reserved forests in the provinces here referred to is upwards of 12,000 square miles or 7,680,000 acres. This sounds a large area, but it is a very small percentage of the aggregate area of the provinces in which these forests are situated.

PANJÁB.

3. A special Forest Regulation exists for the reserved and open forests of Hazára, and the control of the leased forests of Chamba and Bussahir is regulated by rules made under the terms of the lease. The requirements of these forest tracts, therefore, need not be taken into account in the present report.

The total area of demarcated forests in the rest of the province is upwards of 3,500 square miles, the greater part of which (3,300 square miles) consists of the *rakhs* in the Lahore, Rawal Pindi and Multán Divisions. In the greater portion of the demarcated *rakhs* the rights of the State are absolute; in some, however, rights to wood and grazing exist. Eventually it will probably come to this, that in certain districts a large portion of the demarcated *rakhs* will be formed into village forests to provide a regular supply of pasture, selected blocks only being retained. It is obvious that where the rights of Government are not absolute, legal power will be required to effect such a settlement. Nor is it sufficient to provide for the maintenance and improvement of the reserved blocks; power must also be given to regulate the management of the forest lands assigned to villages if it is not intended that these shall be exhausted and made unfit to yield the necessary forest produce to the communities to whom they were assigned.

The rest of the Panjáb reserves which require notice in the present report are mainly in the Hushiarpur and Kangra districts; and here there are, besides the areas demarcated within the last few years in which Government has now acquired complete or almost complete control, considerable extents of forests not specially demarcated in which Government has only limited rights. Legislation is necessary, not only for the reserves, but also in order to ensure the protection and good management of the remaining forests which supply wood and pasture to the agricultural communities of those districts. In Kullu a few limited tracts only have been demarcated, but demarcation of a larger area is contemplated. This larger area will probably not be free of forest rights or privileges, and the rules under which these rights or privileges shall be exercised with the least possible detriment to the forest must be regulated by law.

Excluding the Derajat and Peshawar divisions, the area of the Panjáb is about 80,000 square miles, with a population of 15 millions, and the present Government forests occupy 4·4 per cent. of the total area. But, as already explained, it may be regarded as certain that a large proportion of the *rakhs* will eventually be given up.

Forest legislation in the Panjáb stands as follows:

In May 1855 the Government of India sanctioned "rules for the conservancy of forests and jungles in the hill districts of the Panjáb territories." These rules have the force of law under the Panjáb Laws Act of 1872. The sanction was conveyed in the following terms (paragraph 7 of letter from Secretary to Government of India, Foreign Department, No. 1789, dated 21st May 1855): "His Excellency in Council does not propose to disallow or even alter the general rules which you have submitted for the sanction of Government. His Excellency makes no objection to them as far as they go. But it is very necessary in his opinion that the issue of those rules to the Commissioners should be accompanied by an explanation of the reasons which have led to their being couched in terms so general, and also by directions to each Commissioner to prepare forthwith a set of rules adapted to the peculiar circumstances of his Division." In accordance with these orders it is believed that the Kangra rules of 1859, which are printed on page 91 of Dr. Cleghorn's report on the forests of the Panjáb, were framed ; also rules for the Hushiarpur forests (see Mr. Baden-Powell's Panjáb Forest Report for 1869-70, paragraph 21), and for the forests of Hazára.

The rules of 1855 will be found in the appendix to the present report. In many respects they are sufficient to secure the protection of the forests to which they apply. It will be noticed that they give power, (1st,) to demarcate public preserves and to protect them from trespass, cutting, and interference of all kinds, provided that the authorities do not interfere with the wood or fuel that may be really required by the occupants or owners of the land for agricultural or domestic purposes ; (2nd,) to prohibit absolutely the setting fire to forests, grass, brushwood or other combustible substances (Rule 8) ; (3rd,) to prohibit the grazing of cattle or other domestic animals (Rule 10). The two last-named provisions are evidently not limited to the reserves, but extend to 'all forests required to yield supplies of timber or fuel. Rule 11 makes owners and possessors of cattle responsible for cattle trespass ; and Rule 9 provides that villagers, owners and occupants of the land may be rendered responsible for conflagrations occurring within their bounds. Rule 12 prescribes penalties ; and Rule 13 defines the powers and responsibility of forest officers.

The area to which these rules apply is not clearly defined, but they may be held to be applicable to the hill forests in the Kangra, Hushiarpur, Gurdaspur, Rawal Pindi and Hazára districts. They do not apply to the *rakhs* in the Plains or on the Salt Range. Most of the provisions contained in them it is proposed to incorporate within the new Forest Act. But the Act may possibly not go as far as the rules, and when passed it will have to be considered whether, instead of repealing the rules, it will not be preferable to define the area to which they shall apply, and within that area to leave them in force (with subsidary authority) with regard to all matters which may not be provided for in the Act. It would not do to exclude the area to which they apply from the operation of the new Forest Law, for they are not complete and make no provision for numerous subjects for which provision is needed.

In March 1871 rules were sanctioned and promulgated, under Act VII of 1865, to regulate the use of streams and canals for the floating of timber, collection of drift, unclaimed and stranded timber, and the transit of timber in the Panjab. These rules relate to all timber, but the Government Forest Act is limited to Government timber. In March 1873 rules were sanctioned by the Governor General in Council under the Government Forest Act for the Rawal Pindi forests. In accordance with the Act it is expressly provided in the rules that nothing contained in them shall in any wise abridge or affect any existing rights ; and yet Rule 1 provides that the selected portions of the reserves shall be closed absolutely against all forest rights. These two sets of rules, which were drawn up with great care, illustrate some of the defects of Act VII of 1865. For the Hazara forests, as already mentioned, a special regulation was passed by the Government of India and published on 25th February 1873. That regulation was based upon the Rules for Hill Forests of 1855, and promulgated with reference to 33 Vic., Cap. 3, Section 1. In the first instance it was sanctioned only until 1st April 1874, but by subsequent notification it was continued in force until expressly repealed. This regulation stands independent of the Government Forest Act; it provides for all that is wanted, and the operation of the new Forest Law need not be extended to Hazara. No special forest rules exist for the *rakhs* and plantations in the Plains or for the Salt Range *rakhs*, nor for the Kalesar *sál* forest in the Umballa district.

NORTH-WEST PROVINCES.

4. In Pargana Jaunsar are demarcated 24 square miles of first class forests in which, under the settlement, the rights of the State are absolute, and 119 square miles of second class forests in which the agricultural population of the pargana exercise certain rights to wood and pasture under rules and restrictions laid down in the settlement records. Both classes of forest are demarcated. In the Dehra Dún district there are 285 square miles of demarcated Government forest, 53 square miles of which have been closed by the orders of the Local Government. In the remainder the inhabitants of old established villages in the vicinity graze their cattle against payment, and likewise obtain grass and firewood on payment of fixed rates. In the Saharanpur district 245 square miles of Government forest are demarcated, and it is not finally settled whether they are burdened with rights to wood and pasture. In the same position are 86 square miles in Bijnour, and 137 square miles of demarcated forests in the Bareilly district. In Gorakhpur 115 square miles are demarcated. No free rights, but the villagers near the forests have to be provided against payment with timber, firewood and thatching grass: 211 square miles are demarcated in the Jhansi division. In Kumaun about 500 square miles in the Bhabur and outer hills not yet completely demarcated, the western portion free of rights except in the vicinity of a few villages, and 70 square miles near Ranikhet in which the villagers have the right to collect fallen wood for fuel, grass for their cattle, and to graze where they please, except in portions fenced for reproduction. They have also the right to building timber under certain restrictions. In Garhwál 6 . . . . are miles in the Dúns, outer hills and the Bhabur, not completely demarcated, the greater part free of all rights.

Thus, in the North-West Provinces there are about 2,400 square miles of Government forests the demarcation of which has either been completed or is in course of completion. The total area of the North-West Provinces is 80,900 square miles. The present Government forests, therefore, occupy 2·9 per cent. of the whole area. The acreage is 1,536,000 acres on a population of 31 millions. Draft forest rules framed under the Government Forest Act of 1865 are under consideration and will be found in the appendix. The leased forests in Native Garhwál require no notice here. Extensive forests not separately demarcated are in the hills of British Garhwál and Kumaun, under the control of the district officers; and in parts of these districts the formation of village forests to provide permanently for the requirements of the agricultural population will eventually be found necessary. For such forests the present Act does not provide. Nor does it give sufficient power to protect the system of working the Sub-Himalayan forests of the North-West Provinces which has been in force for many years. Dry wood, bamboos and other forest produce are exported by purchasers, who enter the forests with the permission of the officer in charge and cut and collect timber, bamboos and other forest produce under certain rules and restrictions. Government is the owner of the produce which they export until the purchase-money or royalty, commonly called duty, has been paid. This is either done in advance at the Forest Office or at certain stations on the edge of the forest at the time the produce is exported. Hitherto there have been no difficulties because the people do not know better. But the result of legal proceedings against persons acquainted with the defects of the law who may have succeeded in evading the duty stations and got safe outside the forests with their cart-loads of wood or bamboos would be doubtful under the present Act.

OUDH.

5. In this Province the demarcated forests occupy 823 square miles, equal to 526,720 acres, or 3·4 per cent. of an area of 24,000 square miles with 11 millions of people. With the exception of about 150 square miles which are at the complete disposal of Government, being resumed grants of waste lands, the forests are burdened with rights of wood and grazing which have hitherto greatly impeded all attempts at improving them.

The Oudh Forest rules were promulgated in 1866 under the Government
Forest Act. They were framed on the supposition that the Government forests
were free of grazing rights (Rule XII). But this has been called into question,
and it is certain that portions of the forests must be assigned for pasture.
Owing to the taluqdári tenure the formation of village forests in Oudh is not
practicable, and the tracts that will be assigned for pasture must therefore
remain in the hands of Government.

CENTRAL PROVINCES.

6. The demarcated forests measure 2,270 square miles, or about 2·7 per
cent. of the total area (84,000 square miles). This is 1,452,800 acres on a
population of 9 millions. This area is completely at the disposal of Government and free of all rights of other persons. Besides these reserves, upwards
of 17,200 square miles of excess waste are at the disposal of Government, commonly designated as unreserved forests as they are not reserved from sale and
are available for the extension of cultivation. Forest rules providing for the
protection and management of both reserved and unreserved forests were promulgated in August 1865 under the Government Forest Act of 1865. In these
rules all waste lands which are not private property are classed as Government forests; and this wide definition makes it possible to provide throughout the Government waste lands for the protection of reserved trees, to prohibit dhya cultivation and the cutting of trees and bamboos near springs and
hill streams. It may, however, be doubted whether so wide a definition is
warranted by the Government Forest Act.

COORG.

7. In Coorg the area of the demarcated reserves is larger in proportion
to the total area than in any other province under the Government of India.
They occupy 375 square miles, which is 18·7 per cent. of 2,000 square miles,
the estimated area of the Province. These reserves are believed to be entirely
at the disposal of Government, and their protection is ensured by rules promulgated in August 1871 under the Government Forest Act of 1865. These rules
only apply to demarcated forests enumerated in a separate Notification of the
prescribed form. But in addition to the Government forests the protection of
a number of groves and small forest tracts outside the State forests has been
undertaken by Government. These are the *devara kadus* or sacred groves,
822 in number, the larger proportion measuring under 10 acres, but a few as
large as 50 to 100 acres. These groves are found scattered in all six talukas of
Coorg. They are not temple forests, but have been held in veneration by the
people from time immemorial. Latterly, however, they have been encroached
upon. Many of them have been turned into coffee plantations; and it was
deemed advisable to place them under protection. Rules have been promulgated prohibiting any injury or cutting of the trees in them and interference
of any kind. But as these groves are not Government forests in the sense of
the Act, the rules made for their protection cannot be legalised under its provisions. In most cases these groves are demarcated by trenches and boundary
stones. It is intended to complete their demarcation; and they will thus come
under the head of demarcated public forests and their protection may then be
provided under the forest law proposed in the present report.

BENGAL and ASSAM.

8. In Bengal demarcation is in active progress and no definite information
of the present area of demarcated forests can be given. Rules were promulgated in February 1871 under the Forest Act of 1865. These, like those of
the Central Provinces, apply to two classes of forest—*reserves* which are
demarcated by substantial marks, and *open forests* which are not so demarcated. The area of the reserves was only 120 square miles at the close of
1873-74; but it is understood that, including the reserves of Chittagong and

Palamow and the Sundri Forests of Jessore, it will shortly be brought up to about 2,200 square miles.

In Assam the reserves are reported to cover 615 square miles. Forest administration, however, in that Province, as in Bengal, is still in course of development, and it is not possible to say to what extent the demarcated public forests may be increased. A certain control is exercised also over the forest lands not included within reserves. Forest rules under the Forest Act of 1865, in all respects similar to the Bengal Forest rules and providing for both reserved and open forests, are now under consideration.

BURMA.

9. In Burma the demarcation of the reserves has not yet made much progress. In April 1874 the aggregate area was 279 square miles, but it is believed that during 1874-75 a considerable extent of forest was added to this area. The general principles by which the formation of reserves shall be guided have not yet been settled. But so much is evident that outside the demarcated reserves but within certain geographical limits Government must continue to exercise control over teak and a few other reserved trees, and that the existing rules which regulate the use of streams and canals for the floating of timber, as well as the control over that timber in transit and the management of drift and unclaimed timber, must in some important points be amended and that the area to which they apply must be considerably extended. The Burma Forest rules were promulgated in August 1865 under the Govern-Forest Act of that year, but though the Act had been framed with the view of giving legal effect to these rules, they were not wholly covered by its provisions. The Government Forest Act only provided for control in transit of timber the produce of Government forests. The legality of the rules, therefore, was doubted in respect of timber imported from beyond the frontier, and for this reason the rules were legalised by Act VII of 1869 as regards all timber to which they purport to relate.

Subsequently in 1873 the law on the subject of duty leviable on foreign timber imported into Burma was consolidated by Act XIII of 1873, and the provisions of this Act came in the place of such provisions of the rules legalised in 1869 as related to timber duty.

DEFECTS OF THE PRESENT LAW IN BURMA.

10. The main difficulty in connection with the forest rules in Burma is that the notification defining the area to which the forest rules apply was incorporated in the rules and that this area cannot now be extended without promulgating a new set of rules under the Act. The penalties prescribed by Rule 35 are fine *or* imprisonment and not, as prescribed in the Act, imprisonment only in default of payment. Other difficulties in connection with the present timber and river rules render the revision of the forest law necessary. The most important defect is that the principle laid down in Rule 26 is not supported by the Act. This principle is that all drift and unclaimed teak timber will be considered the property of Government unless proof of ownership is given. Ever since 1856 this principle has been the basis of the existing system of control over timber in transit and drift timber, which was established in order to protect the Government forests and the timber trade of Burma. The formal recognition of this principle is indispensable. The penalties provided by the Act are inadequate; imprisonment is necessary, and penalties *must* be prescribed in addition to confiscation. The Act leaves no option. The Magistrate *must* order the confiscation of implements, that is, axes, marking hammers and similar articles of trifling value or of timber, which in many cases is the property of Government. Having passed this order, he is precluded from imposing any further penalty. There is also a doubt whether the ordinary Magistrates of the Province (except the chief officer charged with the executive administration of a district in criminal matters) have power to act under the rules. From these remarks it will be apparent that for Burma a revision of

the law is necessary. True, most provisions of the existing forest and timber rules of Burma are legalised by Acts VII of 1869 and XIII of 1873. But the area to which these rules apply, cannot be extended and there is no power to amend them.

GENERAL.

11. From the preceding remarks it will be evident that, allowing for further demarcation of reserves, principally in Bengal and Burma, the aggregate area of demarcated Government forests in the provinces mentioned, for the protection of which legislation is considered necessary, now probably amounts to 12,000 square miles, and may, taking into account the land which must be given up in compensation for forest rights or to provide for the wood and pasture requirements of the agricultural population, five years hence be increased to 14,000 square miles. This, as explained before, does not include the demarcated forest tracts in Ajmere and Hazara nor the leased forests, nor those in Mysore and Berar—all of which are provided for by separate forest rules. The demarcated reserves in Hazara, cover 236, those in Ajmere 90, those in Mysore 450, and those in Berar 632 square miles. Nor does the area mentioned include the demarcated forests of Bombay and Madras, though perhaps eventually the Governments of those presidencies may deem it expedient to extend the revised forest law now proposed to the territories under their control.

As regards forest not demarcated the forest law must provide for the few un-demarcated forests in the Panjáb over which Government has forest rights, for the forest lands in the inner hills of Garhwal and Kumaun, for the unreserved forests of the Central Provinces, for the open forests of Bengal and Assam, and for a large extent of forest in Burma which will not be included in the demarcated reserves.

With the view of further illustrating these introductory remarks and in order to facilitate reference to those documents, the Government Forest Act of 1865, as well as the rules made under it and the Acts relating to the forest rules and timber duty in Burma, have been collected in an Appendix to the present Report; and in Burma the new draft forest and timber rules framed by Mr. Baden-Powell while on deputation to that province have been included. The forest rules and regulations, made for the provinces and districts to which the Forest Act has not been extended have been added in order to complete the collection. It has been said that most of the existing forest rules are useless. I do not agree with that view. Experience has shown that they are useful as far as they go; but they are insufficient and defective, and a fresh forest law is required to remedy their defects. At the end of the Appendix will be found the revised draft Forest Bill framed by Mr. Baden-Powell in 1874 when officiating as Inspector General of Forests.

FIRST CHAPTER.
OF DEMARCATED PUBLIC FORESTS.

12. *State and village forests.*—With the exception of the sacred groves of Coorg, such forests only have hitherto been demarcated in the provinces to which the forest law, if passed; will be applicable, as were the property of the State and were set apart for State purposes. But it is certain that in those districts and provinces where the old village organisation and communal village lands have been maintained forest lands will also eventually come to be demarcated for the use and benefit of villages or groups of villages, and it is equally certain that those village or communal forests will require the protection of a special law as much as the State forests. So also it may in exceptional cases, as in Coorg, be considered desirable to have the power to protect by law forests

which have hitherto been protected because they were regarded as sacred. In such cases it will probably be found desirable to demarcate the tracts intended to be protected in order distinctly to limit and define the responsibility which Government undertakes in a matter with which, as a rule, it does not concern itself. The Forest law therefore should give power to make rules for the demarcation, protection and management of village forests, as well as of State forests. The right to control village forests will be based in most cases upon the fact that Government is the proprietor of the land assigned to the village, and in other cases upon the rights which Government has in the lands of the village. The rules required for village forests will be similar to those required for State forests. Both State and village forests may be comprised within the term public forests, but the shorter and more usual term—*reserve* or *reserved forest*—will also occasionally be used in this report. The first point then for consideration will be how far legislation is required to ensure the protection and good management of demarcated public forests.

13. *Procedure in demarcating forests.*—The demarcation of these forests is necessarily a one-sided act on the part of Government. The practice hitherto has been for the local civil and forest officers, acting jointly, to consider the interests and requirements of the surrounding population, and with due regard to those interests and requirements, to fix the boundaries of the forest and to demarcate and notify them. In some cases, as in Oudh, a special demarcating officer was entrusted with the duty. In these demarcated forests the Government exercises proprietary rights, in some cases without limitation, in others subject to rights and privileges of the neighbouring population. For those forests which have not yet been demarcated, the procedure of demarcating (or constituting) the forest must be prescribed, and this may be done by means of rules to be made by Local Governments under the sanction of the Government of India. The Act, however, should prescribe the following :—

(a.)—The constitution of a public demarcated forest must be effected by a settlement or civil officer to be styled the demarcating officer, acting in communication with a forest officer, whose proposals and objections must be recorded and duly considered.

(b.)—The officer charged with the constitution of a public demarcated forest shall, in making his proposals, consider, *first*, the convenience and well-being of the people resident in, or in the vicinity of, the forest; *second*, the present and probable future demand of the district in timber, wood, bamboos, grass, and other forest produce; *third*, the utility of the forest in preserving the soil, the water-supply in springs and rivers, and the general effect of the forest on the fertility of the lands in its vicinity.

(c.)—The proposals of the demarcating officer must be subject to the confirmation of the Local Government.

(d.)—If the boundaries of any demarcated public forest are not clearly and unmistakeably indicated by roads, rivers, or other existing boundaries or land-marks, they must be marked out by permanent marks in such manner as the Local Government may direct.

The demarcation of a forest is a proceeding analogous to the selection of waste lands to be sold or otherwise disposed of for cultivation. Under Act XXIII of 1863 the duty of deciding whether any land may properly be sold as waste land is entrusted to the Collector. The demarcation of reserves is also analogous to the demarcation of excess waste, and it will be useful here to refer to the sections on that subject of the North-Western Provinces Land Revenue Act XIX of 1873. The procedure for demarcating the excess waste is laid down in Sections 57 to 60, and Section 60 gives power to the settlement officer, if the proprietor of the adjoining mahal proves that he has theretofore enjoyed the use of the waste land for pastural or agricultural purposes, to assign to such mahal so much of such waste land as he may consider requisite for such purposes; and he shall mark off the remainder and declare it to be the property of Government. This is exactly what is required in the demarcation of Government forests. The demarcating officer is to assign to the adjoining villages so much of the waste land as he may consider requisite for their pastural or agricultural purposes, and shall mark off the remainder and declare it to be the property of Government. This duty must obviously be entrusted to an officer familiar with the husbandry and pastural requirements of the people who resort to the forest, and for this reason the responsibility of deciding, what boundaries to select, must rest with a civil or settlement officer specially selected for that purpose. But the proposals regarding the boundaries to be selected must be prepared and laid before him by the forest officer, and both the original proposals of the forest officer, as well as any objection which he may raise to the proceedings of the demarcating officer, must be recorded. In case of a difference of opinion between the forest and demarcating officer, this provision will give the Local Government an opportunity of ordering a revision of the demarcation, and in the case of bad work it will make it clear who is responsible for it. This arrangement will ensure prompt action, and at the same time Government will have the full benefit of the forest officer's professional knowledge and experience. At first sight it may not appear to be very important whether this arrangement is adopted, or that suggested by Mr. Baden-Powell, to entrust the work to a commission of two officers—one a forest and the other a civil or settlement officer,—but clear and undivided responsibility is a great advantage, provided due care is taken to place the contrary opinion of the Forest Department on record. The main extension of the State forest area within the next ten years may, if all goes well, be expected in the Eastern Provinces, Bengal, Assam, and Burma, and the bulk of the work ought to be completed within the next five years. For that time it will probably be found convenient in those provinces, where heavy demarcation work is in progress, to appoint special officers for this work.

In Burma, for instance, as soon as the general question of the classification of forests has been finally settled, it will probably be expedient to employ several forest officers to select and mark off in a preliminary manner the boundaries of State forests, so as to have all proposals ready when the demarcating officer comes round, who will frame and submit his final proposals to Government.

After the demarcation of the State forests has been completed, there will, in those districts where the old village organization has been maintained, remain the task of forming and demarcating village forests. But it would be premature now to consider the arrangements which this further measure will demand, the necessity of which has not yet been recognized. For village forests it must at present suffice to lay down the general principle, that the rules for their demarcation, protection and management will be similar to those proposed for State forests.

14. *Forest demarcation aims at a separation of rights.*—Obviously the first point to be established is, to authorize Local Governments to demarcate forests and to declare them to be either State or village forests. In giving this authority the rights of others must be guarded, and this is the most difficult part of the whole business. Act VII of 1865 simply said that the notification of a Government forest shall not abridge or affect any existing rights of individuals or communities. This, however, does not in the least advance the settlement of the question; it leaves the position of Government in the Government forests precisely as it was before. The practical object in all cases of demarcating public forests is to effect a separation of rights, the result being the acquisition of more extensive rights by Government within the demarcated area, against concessions made outside its limits to the surrounding population. The ultimate effect of this separation of rights will be beneficial to both parties; the right-holders will obtain more complete control over an area smaller in extent but conveniently situated for the supply of their wants, while, on the other hand, the protection and good management of the State

forests, which can only be secured by means of this separation of rights, will, eventually confer incalculable benefits upon the country at large, including the right-holders themselves. To ensure the attainment of these ends by regulating the separation of rights here adverted to in a just and equitable manner, is a legitimate and important task to be undertaken by the Legislature of India.

15. *Legal effect of the demarcation of a public forest.*—It will be necessary to provide that the following consequences shall ensue on the demarcation of any public forest:—

(*a.*)—No fresh rights of user or easement shall accrue or be acquired in any demarcated public forest.

(*b.*)—No portion of such forest shall be alienated by sale or lease without the sanction of the Government of India, and any sale or lease made without such sanction shall be null and void. Every authorized alienation of a State forest shall be notified in the official *Gazette.**

(*c.*)—Privileges which may be granted in any public demarcated forest for the convenience of the people (such as grazing, the use of timber, wood, or other forest produce) shall be for the use of the land or premises, or of the person to whom they are granted, but not for sale, lease, or merchandize, and no produce obtained in virtue of such a privilege shall be sold or bartered. All such privileges shall be granted subject to the condition that they shall be exercised only in such portions of the forest as may be assigned for that purpose by competent authority.

(*d.*)—The Local Government shall have the power to determine what roads and pathways shall be authorized for public traffic, to cause all other roads and pathways to be closed, and to prohibit all ingress into the forest except on the authorized roads and pathways. Provided that the authorized roads and pathways must be sufficient for the requirements of the existing traffic, and that any persons or communities who, within the time fixed and in the manner prescribed shall have established a right of way other than on the authorized roads or pathways, shall receive full compensation for the loss of such right.

(*e.*)—The Local Government shall have the power to prohibit any kind of cultivation or clearings for the purpose of cultivation in a demarcated forest. Provided that no proprietary or occupancy rights are interfered with by such order.

As a necessary consequence of these provisions, the manner in which and the time within which claims to rights of way and proprietary or occupancy rights in the land shall be preferred, must be fixed. The decision of claims arising out of these matters, unless admitted by the demarcating officer, may be left to the ordinary civil or revenue courts of the country. Rights of way, of occupancy, or proprietary rights in the forests have nothing peculiar which would require a special agency, and when such rights are established, and it is found necessary to extinguish them, then the compensation to be paid may be settled in accordance with the provisions of the Land Acquisition Act.

The consequences of demarcation will therefore not interfere with untained, rights. Nevertheless the alterations which, as here proposed, the demar in which effect in the status of the demarcated area, are very important. No fresh xtent and accrue; the land will not be alienable except under sanction of Goverr the great those who have private rights of way or proprietary or occupancy rights enclosed in the reserve must within a fixed time and in a prescribed m g wood and their claims, and, unless they are admitted by the demarcating offi le of a right them within a given time in the Courts which have ordinary jurisdi e Conference ion, the right

* This provision is essential, for it will be a long time before the necessity of maintaining inta laiming title sated forests is fully recognised in this country, years (Section

matter. No interference however with rights to wood and pasture, or any of the other forest rights which may exist, is contemplated.

16. *Rights of common in England and forest rights as understood in France and Germany.*—It is a question open to discussion whether these provisions might not usefully and equitably go a step further. The rights to pasture, wood, and other forest produce, which the agricultural population in the vicinity of our Indian forests have exercised from time immemorial, have the character of rights of common. "It is the right which one or more persons may have to take or use some portion of that which another person's soil naturally produces;" and if we enquire into the origin of these rights, it will be found that in most cases they may be compared to what is called "the right of common appendant," by which the owner or occupier of arable land, or, in the case of commons of estovers and turbary, the owner or occupier of a house is entitled to the use of the manorial waste for such purposes as are necessary to the maintenance of his husbandry or premises.

Now, as regards the common of pasture, it is I believe agreed that the right of common appendant is limited to such cattle as the land can maintain during the winter by its produce, or requires to plough and compester (manure) it. The commoner cannot pasture the cattle of a stranger for hire ; he cannot rent his right.

So also the common of estovers, which includes the right to cut or procure from the forest, or other wastes of another, wood for his building (house-bote), inclosing and firing or other necessary purposes (plough-bote, cart-bote, and hay-bote), is governed by rules analogous to those explained. "If there be a prescription for fire-bote to burn in a hall, this will not extend to the consumption upon the same premises after they have been converted into a kitchen or a malt-house," and "the estovers taken must be spent upon the premises which give the right to take them." Thus also turf dug upon another man's ground in virtue of common of turbary must be expended on the premises to which the right is appendant. It is, I believe, admitted that in his user of the common the commoner is limited to his reasonable exigencies for the purposes for which the rights are conferred. Thus, the pasture must be for his own cattle ; the fire-bote, a reasonable quantity for the fuel of the chimneys to which it is attached ; the estovers, a reasonable quantity for the repair of his own fences or farm implements.

The same principles are recognised in the French and German Forest laws. It will suffice to invite attention to the following articles of the *Code Forestier* :—

Art. 70. *Les usagers ne pourront jouir de leurs droits de pâturage et de panage que pour les bestiaux à leur propre usage, et non pour ceux dont ils font commerce.*

Art. 83. *Il est interdit aux usagers de vendre ou d'échanger les bois qui leur sont délivrés et de les employer à aucune autre destination que celle pour laquelle le droit d'usage a été accordé.*

The preceding remarks will explain that in Europe rights to wood and pasture in the forest of another are, as a rule, and unless there is an agreement between the persons concerned to the contrary, subject to certain limitations arising out of the nature of these rights. This limitation is not arbitrary ; it follows as a necessary consequence from the origin and development of these rights. And whenever the subject has been enquired into with care in this country, the principle h̲a̲s̲ been recognised. Thus the settlement of rights in Jaunsar prohibits the of a Gov or exchange of any rights in the second class forests, and prohibits individual wood cut in those forests by virtue of a right. And in Section 27 settlement ft Bill of 1869 it was proposed that this limitation of forest rights ment fores sanctioned by law.

demarcatin ontinental Forest laws recognise a further limitation of customary the acquisit in another respect. Thus Article 65 of the *Code Forestier* provides area, agains ice des droits d'usage pourra toujours être réduit par l'Administration, The ultimat et la possibilité des forêts."—that is, the exercise of those forest rights parties ; the ed in accordance with the condition and the possible yield of the smaller in e: regard to pastures, this is defined in Articles 67 to 69, which while, on the stant toutes possessions contraires) pasture in any blocks which have not 'been declared *défensables*, i. e., safe against cattle, by the forest administration. The blocks open for the admission of cattle and the number of cattle which each right-holder may send to graze are fixed by the forest administration. The principle upon which these provisions are based is, as regards wood, that right-holders can not claim more in one year than the normal permanent annual yield of the forest; and as regards pasture, that the exercise of the right must not diminish the productiveness of the forest in wood or timber. Similarly in the Jaunsar settlement paper the forest officer is authorised to close certain blocks, provided sufficient forest is left open for the requirements of the people,

17. *Is the customary use of the forests in India based on right or on privilege?* —In Sections 25 and 26 of the draft Forest Bill of 1869 it was proposed to adopt these principles with regard to forest rights in India, and accordingly to sanction the closing of such portions of any demarcated forest as might be deemed necessary for the protection and improvement of that forest. However when the draft Forest Bill was circulated these proposals were strongly objected to as an attempt to curtail and restrict the exercise of private rights. In reality the proposal was to give expression to that limitation of forest rights which is inherent in their nature and which follows as a necessary consequence from their origin and development.

On the other hand, it has been maintained that the customary use of the forest in India here spoken of is usually not based upon a right but upon a privilege. Great stress has been laid upon this point in Mr. Baden-Powell's able paper on forest legislation in the Report on the Conference of 1574. He argues that the villagers who from time immemorial were accustomed to cut and graze in the nearest jungle lands did not acquire a right by prescription, because they used the forest without any distinct grant or license and without any idea of asserting a right as against the ruling power or against other individuals or communities ; that the State had not exercised its full right over the forests, which were left open to any one who chose to use them ; but that the right of the State was unimpaired and was asserted whenever a Native Ruler chose to close whole areas of forest to preserve the game, as in the well-known instances of the Belas of Sindh enclosed by the Amirs.

This view of the case merits careful consideration, and doubtless in many cases what are sometimes called forest rights are not rights at all, but merely privileges which are exercised by permission and at the pleasure of Government and not as of right. A large class of cases will, however, remain and must be provided for by the Forest law, in which the custom to graze the village cattle and to cut wood for the requirements of the village have grown up in a manner in every respect similar to the growth of rights of common or of forest rights in Europe. The fact that the former Rulers in many cases have extinguished such customary user of the forest in a summary manner and without compensation is hardly an argument in point, for those were cases of might against right. As against other individuals and communities the customary rights to wood and pasture have as a rule been strenuously maintained. Mr. Baden-Powell on page 4 of the Conference Report classes the great mass of the forests in the provinces and districts to which this report applies as State forests in which no rights exist, and limits the forests in which the property of the State is burdened with real legal rights to those in which Government have granted recognised, though undefined, rights properly so-called by solemn record at settlement. If this view of the case could be maintained, forest legislation in India would be a very easy task, for the forests in which Government have granted specific rights at settlement are limited in extent and are not numerous. Act VII of 1865 would in that case be sufficient for the great mass of Government forests.

The question whether the customary practices of grazing, cutting wood and collecting forest produce in any particular forest are exercised by virtue of a right or a privilege is not as a rule an easy one to decide. On page 6 of the Conference Report Mr. Baden-Powell states that to constitute a right by prescription, the right should have been peaceably and openly enjoyed by any person claiming title thereto as an easement and as of right without interruption for 20 years (Section

27 of Act IX of 1871). This test relates to easements which are not ordinarily
held to intlude the forest rights here referred to. But even accepting this test
as applicable to forest rights, there are numerous forest tracts in which the
inhabitants of the vicinity have exercised these rights peaceably and openly
without interruption for more than 20 years. Whether these people have ever
made it clear to themselves that they grazed their cattle as a right will in most
cases be impossible to determine. I have always held that the growth of forest
rights in India has been analogous to the growth of similar rights of user in
Europe. There are many well known cases in which forest rights in Europe have
arisen out of a specific grant, and in such cases the extent of the right is construed
by the terms of the grant and is not necessarily restricted by the limitations
adverted to. In most instances, however, they have grown up out of the use
by the surrounding villages of the common waste and forest. Forest rights in
India have had a similar origin and development as in Europe, with that important
difference that the arbitrary dealings of the Native Rulers have interfered with
the growth of these rights and have in many cases restricted or extinguished
them. It is obviously impossible to come to a final decision at present on the
general question raised by Mr. Baden-Powell. Meanwhile forest legislation should
sanction a mode of procedure by which the exercise of rights injurious to the forest
may with due regard to the requirements and convenience of the people who now have
a beneficial interest in the use of the forest be regulated, restricted or extinguished;
and this procedure will à *fortiori* apply to those forests in which these customary
practices may hereafter be found to be based on privilege only and not on right.
The sections of the draft Bill of 1869 adverted to above were intended to
apply to forest rights and not to privileges; and it is for consideration whether the
new forest law shall not provide that, unless a right to the contrary is established,
customary rights to pasture, wood and other produce of a demarcated public forest
shall be exercised subject to the following limitations: 1*st*, the exercise of
customary forest rights shall be limited to the material required for the lands or
houses in which or persons in whom the right is vested and not for sale or barter;
2*nd*, customary forest rights shall be exercised only in such portions of the
forest as may from time to time be assigned for that purpose under the authority
of the Local Government, provided that the area thus assigned is sufficient to meet
the ordinary requirements of the right-holders.

The first of these two provisions if accepted may be inserted along with the
other provisions explained in paragraph 15, whereas the second may perhaps find
a more appropriate place further on.

18. *Authority to determine and record existing forest rights.*—Under all circumstances power must be given to the Local Government to take the needful
steps for ascertaining, determining and recording the existing forest rights of other
persons in a public demarcated forest. The law must lay down the procedure for
deciding claims to forest rights, and it must preclude the recognition of any rights
which have not been claimed and admitted by competent authority within a certain
period commencing from the date of the notification inviting claims to forest
rights. With regard to this matter we must keep in mind that it is the nature
of forest rights to be obscure and undefined,—that the act of demarcating a forest
out of the mass of waste and forest land hitherto open to the surrounding population and used by them under certain restrictions introduces a new feature, and
that it cannot in all cases be expected that the people interested in the use of the
forest will readily make up their mind as to the rights which they actually possess.
There are districts, particularly in the North-West Himalaya, where forest rights
to wood and pasture are well understood and defined; and in all such cases the
nature and extent of these rights in demarcated forests should be formally determined and recorded at the time when the forest is demarcated, or in the case of
forests already demarcated within a short time after the promulgation of the rules
made under this Act. But no useful object would be gained in insisting on the
demarcation of forest rights in all demarcated public forests. Thus in the reserves
of the Central Provinces and in a portion of the Garhwal and Kumaun sâl forests
it may be assumed as certain that the rights of the State are absolute and that no

difficulties will be felt in that respect. Most of these forests have been under the undisputed control of the Forest Department for a series of years, and it would only lead to embarrassment if claims to forest rights were invited. In the Central Provinces the malguzars of the vicinity of the reserves would come and revive their claims to an extension of their village area; and in the Patli-Dún and adjoining forests of Garhwál and Kumaun numerous claims would be raised afresh which were settled long ago when these forests were first constituted. In such and many similar cases it will certainly be found preferable to leave matters *in statu quo* and to take the chance of objections being raised in the usual course, Nor will it be necessary to invite claims where the rights in demarcated forests have already been defined in the settlement records, as in the first and second class forests of Jaunsar and the forests of the Kangra and Hushiarpur districts. It is essential that the Act should give power to determine existing forest rights, but the determination of forest rights must not be made compulsory.

19. *One-sided record of forest rights and privileges.*—For all public demarcated forests, however, a record must be drawn up of the forest rights which in the opinion of the Government officers exist in each forest, and it will be convenient to include in this record a statement of all privileges not of the nature of rights which have been conceded by Government. These records, however useful (if prepared with due care), will be one-sided statements; and it is for consideration whether the Forest Act need say any thing regarding them. The needful orders for the preparation of such records have already been issued by the Government of India in a Circular of the 3rd April 1875. It is a purely administrative measure the provisions of which may require to be modified from time to time.

20. *Regulation and extinction of forest rights.*—The next step which must be taken in order to ensure the improvement and efficient control of the forests is to empower Government to regulate the exercise of these rights and in some cases to commute, restrict or extinguish them,—full compensation being given to the estates, communities or persons whose rights may be affected by these proceedings. The procedure which I would suggest should be sanctioned by law will best be explained by stating three cases in which the settlement of forest rights is pending or proposed :

The *first* case which I desire to mention is the settlement of the grazing rights or privileges in the Government forests of the Debra Dún. The people here pay for the right of pasture, and it is uncertain whether they graze their cattle in these forests by right or by privilege. The Local Government have already decided upon the report of the local officers to restrict the pasture to certain portions of the forests sufficient to furnish the needful grazing, and to close the remainder. The Local Government have also authorised the Conservator of Forest, acting in communication with the Superintendent of the Dún, to demarcate the forest tracts which shall remain open for grazing and those which shall be closed. What seems to be required in similar cases is to authorise Local Governments to make such a settlement, to prescribe the procedure by which it is to be effected, and to fix the period of limitation for claims against the arrangements made under these provisions. Should the suggestions of paragraph 17 be adopted, this case may be dealt with under the provisions there proposed.

The *second* case is as follows : In certain demarcated forests of Pargana Jaunsar in the North-West Provinces known under the name of second class forests the landholders of the surrounding villages have certain forest rights of which the right of pasture is the most important. In order to provide a permanent supply of fuel for the station of Chakrata, it is necessary to close an area of about 5,000 acres of the Deoban Forests, which contains the high level grazing ground for the sheep and goats of about 15 villages. This can only be effected by a somewhat complicated system of give-and-take. The summer pastures of these villages must be shifted to a more distant part of the main ridge, and for the inconvenience of the greater distance the people must be compensated by the grant of other land in the vicinity of their villages. Again some compensation must be given to the people who under these arrangements will have to share their high level pastures with the sheep and goats of the 15 villages in question. In the same manner their rights to building wood, branches for fencing, litter and cattle fodder, and other material must be commuted—in some cases by the grant to them of other forest land, in others by the payment of money or the remission of taxes. It is evident that the first proceeding must be to demarcate the area intended to be closed, so that it may be perfectly clear to all persons interested what is proposed to be done, and then for the forest officers, acting in communication with the civil officers, to make the needful proposals to the persons interested for the commutation of their rights. These are operations which must be done on the spot. All objections also to the arrangements proposed should be heard and decided on the spot by the authority appointed for the decision of these matters.

The *third* case relates to the rights and privileges of the villages in the vicinity of the Government forests of Oudh. The principal of these rights are to cut fuel and to graze cattle; and they are exercised over the greater part of the forests. The object which we must aim at in the Oudh forests is to restrict the area over which these rights are exercised, and in lieu of the more extensive area to grant additional rights in the smaller area assigned to the right-holders, such as the permission to cut timber and under certain conditions to break up land for cultivation. Under the talukdari tenure of Oudh the matter is complicated by the necessity of considering both the interests of the talukdars and the ryots in the settlement of this question.

21. *Procedure proposed.*—The proposals for the commutation of forest rights by means of an exchange of land or other compensation must in the first instance be made by the forest officers acting in concert with the local civil officers. If the parties interested consent to these proposals, or if any objection made to them be adjusted by mutual agreement, then the consent or agreement should be recorded, and the forest law should prescribe the form of such record and its value as evidence. If on the other hand no consent or agreement is come to, the decision must be left to another authority. Two different proposals have been made in this respect. The drafts of 1868 (Sections 9 and 10) and of 1869 (Section 33) proposed that commissions be appointed by the Local Government for the decision of these cases. Mr. Powell's Bill (Section 39) proposed to adopt the procedure of the Land Acquisition Act. The peculiar feature of this business is that the number of interested persons is large,—that the arrangements which must be made to satisfy their requirements are complicated, and that in many cases the decision will be not only whether certain proposals shall be admitted or rejected, but that the local officers making the proposal will during the course of the enquiry be called upon to modify them in order to enable the authority with whom the final decision rests to sanction the arrangement. It is essential that matters of this kind should be settled on the spot in the presence of the interested parties and with that perfect knowledge of the details of the case which cannot be acquired without an examination of the locality. The issues submitted to the decision of the Court and Assessors under the Land Acquisition Act are of a much more simple nature and can be determined without necessitating the examination of the locality by the Court and its Assessors. Under these circumstances the only plan that I can suggest is that proposed in the draft Bills of 1868 and 1869, *viz.*, to entrust the decision of these matters either to a single officer specially selected or to a commission,—the selection of the members, mode of procedure, and form of decision being regulated by rules to be framed by Local Governments under the sanction of the Government of India. It should be considered whether the decision of the officer or commission should be final when sanctioned by the Local Government, or whether it should become final after the lapse of two years from the date of publication, unless reversed during that time by a decree of a Court specially designated to hear appeals from such decisions.

22. *Customary practices injurious to forests.*—In the preceding paragraphs all that requires to be said regarding the general question of forest rights and privileges has been explained, and it may now be useful, even at the risk of repeating some of the remarks previously made, to review more in detail the different kinds of customary user of the forests which interfere with their protection and improvement and the manner in which it is proposed to regulate, commute or

extinguish them under the new forest law. Regarding the rights of way, of occupancy of land, and proprietary rights in the forest, the needful has been said in paragraph 15.

The principal other customary practices which are injurious to the forests are the following:—

I.—Pasture of cattle, cutting grass and branches for fodder and litter of cattle.

II.—Setting fire to grass or forest, either to improve the grazing, or for other purposes.

III.—Cutting and removal of timber, wood, bamboo and brushwood, including fuel, charcoal and material for fencing.

IV.—Cutting leaves and branches for manure (which are generally spread over the field and burnt), including cutting of dead leaves and the removal of the surface soil.

V.—Making catechu, collecting mowha, fruits, wax, gum and other minor forest produce.

VI.—Digging stones and *kankar*.

23. *Pasture and fires.*—Of these pasture and fires are by far the most important, and injurious. In Burma, where pasture in the forests is of less moment than in most other provinces, fires are mainly caused by the practice of burning the cut jungle in clearances made for toungya cultivation. This cause of fires can be excluded from the reserves by excluding toungyas where no occupancy rights exist. Other causes of fire can only be eliminated by prohibiting ingress into the reserves during the hot season. Pasture is injurious directly, by the damage which the cattle do to the forest growth, and indirectly by the practice of herdsmen to light fires. In most forests, the first aim must be gradually to exclude cattle and fires from demarcated forests, if not throughout the year, at least during certain seasons. There seems no objection to empower Local Governments to close any demarcated public forest, or any portion of such a forest, against fires and cattle, and to prohibit the cutting and removal of grass, leaves, and branches for litter and cattle fodder, provided (1) no specific rights in that particular block to be closed* are interfered with; (2) the actual requirements of the communities or persons who, at the time of closing the forest, were in the habit of using it for pasture and for the provision of litter and cattle fodder are sufficiently provided for elsewhere. Of this the order of the Local Government closing the forest should be regarded as sufficient evidence.

A provision to this effect would be sufficient to legalise the action taken by the North-West Government in the case of the Dehra Dun, and it would in every respect be beneficial and equitable. It would also authorise the closing of a portion of the Oudh forests against cattle and fires. It would also legalise the closing of a large portion of the Garhwál and Kumaun sál forests, and it would authorise the conversion into a first class forest of some out-of-the-way tracts in Jaunsar, such as the deodar forest on the headwaters of the Riknargad, which, owing to their distance from villages, are not much resorted to for pasture.

This provision, however, would not suffice in the case of some of the Kangra forests in which the Guddees from the higher ranges of Chamba have definite prescriptive rights to graze their sheep during winter. Nor would it suffice, in order to effect the arrangement proposed in paragraph 20, to convert the Deoban forest of Jaunsar into a first class reserve. In such cases, the procedure must be, to come to an agreement with the right-holders; and, failing this, to appoint either one officer, or a commission of officers, for the settlement of the matter. These must hear the proposals of both parties on the spot, and their duty must be to effect a just and equitable arrangement, and to award full compensation for any rights which it may be proposed to restrict or to extinguish. This, it has already been explained, cannot be done by the ordinary Civil Courts, nor can it be done by the agency prescribed by the Land Acquisition Act. It must be done by a special officer, or commission of officers, who can transact their business on the spot,

---

* It is for consideration whether it may not be provided that such specific rights shall be established before the ordinary Civil Courts within two years from the date of the notice closing the forest. The action here suggested will not be taken unless there is a strong presumption that there are no such specific rights.

The compensation to be given in the case of restriction or extinction of this class of rights will consist either in the grant of land or of grazing rights in other Government land, or in the payment of a sum of money or in periodical payments, or in the abandonment of the right of Government in any forest or other land in favor of the person or estate affected, or in the remission or reduction of any taxes or other Government demand, or in any of these conjointly (Mr. Baden-Powell's Bill, Section 37).

The proposals under this head briefly amount to the following : Local Governments should be empowered to act according to the circumstances of the case,—either to close the whole or portions of a reserve, provided the actual grazing requirements of the people are otherwise satisfied, and actual right-holders are not interfered with in the enjoyment of their rights ; or, in cases where adverse rights are known to exist, to obtain the needful authority for closing the reserve,—either by means of an agreement with the right-holders, or, failing this, by a settlement effected under the orders of the officer or commission appointed for the purpose.

24. *User of wood and other forest produce.*—The practices enumerated under heads III to VI are of much less importance ; they do not, as a rule, injure the productiveness of the forest. Unless they are excessive, the damage done by them mainly consists in diminishing the revenue of the forest by the removal of material without payment.

In regulating the exercise of these practices, one of two plans may be followed. Either the practices, whether based upon rights or privileges, may be restricted to a certain area. This the Local Government may be empowered to order in those cases where the actual requirements of the persons and communities can be satisfied from the restricted area, on condition that no specific rights are infringed and restricted. As in the case of pasture, this can generally be done in thinly populated districts ; and the effect of such an order will be this,—that those who can establish a specific right to get wood or other produce from a certain locality will continue to exercise it, provided they prove it within a fixed time in the ordinary Civil Courts, while those who take no steps, or who fail to establish it, will only be permitted to continue their practices within the restricted area.

Thus, in demarcating State forests in the hilly tracts of Pegu, the cutting of wood and bamboos and the cutting of forest produce has hitherto been prohibited within the reserves, on the understanding that ample forest is left outside the reserves for the wants of the population in the vicinity, and that no actual rights are infringed by closing the reserves. In other provinces, also, this procedure has actually been followed ; and the present proposals aim at nothing else than to legalise the present practice, and to give an opportunity to right-holders, who in the preliminary enquiry may not have been discovered, to establish their rights, and to continue to them the enjoyment of these rights.

There are, however, cases the conditions of which do not admit of such summary procedure. In such cases, the first step must be to define the quantity and description of timber, wood, bamboo, brushwood, leaves or other produce of forest or its soil, to a supply of which, annually or at certain seasons, the right-holder is entitled. The rights mentioned under III to VI are capable of being defined in this manner ; but the definition will, in most cases, be found an extremely difficult task, and in such cases it will not be worth while to undertake it except as preliminary to the commutation or extinction of these rights.

The procedure will be for the local forest officer, or for an officer specially charged with this duty, after due enquiry, to prepare a list of right-holders and a statement of the rights to which each is entitled, defined as indicated above. The list and statement should be submitted to the officer or commission proposed to be appointed ; and they should, after giving due notice to all persons interested, hear any objection that may be proposed and finally decide the matter. It is evident that this must be done on the spot, and can neither be done by the ordinary Civil Courts nor by the agency employed for determining the compensation to be paid when land is acquired for public purposes.

As already explained, in such cases it will not entail much additional work to go a step further, and to effect a commutation of these rights. In some instances, this may be effected by assigning to the right-holders such an extent of forest land as may be capable of furnishing the material to which they are entitled. When this is not practicable, compensation of another kind must be given, which may consist in the payment of a sum of money or in periodical payments, or in the abandonment of the right of Government in any forest or other land in favor of the person or estate affected, or in the remission or reduction of any taxes or other Government demand, or in any of these conjointly. The proposals for compensation must, as in the case of defining the forest rights, in the first instance, be proposed by the local forest officer, or other officer specially appointed in that behalf, and they must then be communicated to the right-holders, who, if they do not agree, should appeal to the officer or commission with whom the final decision will rest.

25. *Conclusion.*—This is all that in my opinion requires legislation in respect of demarcated public forests. It amounts to this—

I.—Local Governments must be empowered to demarcate and notify public forests and be instructed to effect this by means of an officer specially appointed for the purpose, to be styled the demarcating officer. The legal effects of the notice notifying the demarcation will be—(*a*) that no new prescriptive rights shall accrue; (*b*) that the land cannot be alienated without the sanction of the Government of India; (*c*) that forest privileges shall only be granted subject to certain restrictions; (*d*) that the use of roads may be regulated, and ingress, except on authorized roads, may be prohibited; (*e*) that all cultivation may be prohibited except by those who have proprietary or occupancy rights.

II.—Local Governments must be empowered to close demarcated public forests, or certain portions of them, against pasture and the user of wood or other forest produce, provided that persons who within a fixed time establish their rights are permitted to continue the exercise of these rights, and provided the requirements of the persons, or communities who have hitherto been in the habit of using the forest, are fully provided for in the open portions or elsewhere.

III.—Local Governments must be empowered to appoint an officer, or a commission of officers, for the purpose of defining forest rights, and, when necessary, for commuting and extinguishing them against full and equitable compensation to be fixed either by agreement with the interested parties, or by the said officer or commission appointed.

IV.—Local Governments must be authorized, under the sanction of the Governor General in Council, to promulgate rules for the protection and management of demarcated public forests, and to impose penalties for the breach of such rules within certain limits.

These rules may prohibit the following acts :—

(1.) Setting fire to the grass or forest, or kindling any fire in it or in the vicinity thereof, without effectually preventing its spread into the forest.

(2.) Burning lime or charcoal without permission.

(3.) Trespass by men or cattle off the authorized roads and pathways.

(4.) Grazing or pasturing of cattle, except with the permission of the Conservator of Forests.

(5.) Felling, girdling, cutting or lopping, marking, burning, stripping off bark or leaves, tapping for gum or resin, or otherwise injuring any trees, shrubs, or bamboos, except with the permission of the Conservator of Forests.

(6.) Removal of dead leaves, turf, or the surface of the soil, cutting grass, collecting fruits, honey, wax, bark, gum, lac, or any kind of forest produce, without the permission of the Conservator of Forests.

(7.) Temporary clearings or every other form of cultivation without the permission of the Conservator of Forests.

These rules should regulate the disposal of timber and other produce of the forests, and they may also authorize the levying of tolls on roads and bridges built within the forests.

It should be condidered, whether the Penal Code sufficiently provides against moving and injury of boundary marks of the forest and of blocks and compartments. Should this be doubtful, then a section prohibiting all interference with boundary marks under penalties should be added.

Such rules would not affect or abridge any existing rights which may have been established within a fixed time and in the manner prescribed.

It appears to me questionable whether it is necessary to make special provision for cases of disputed boundaries. At the time of demarcation, the boundaries will be laid down by the demarcating officer, and special provisions for boundary disputes, which may arise at a future time, are not urgently required. Legislation on this point, therefore, may be deferred for the present.

SECOND CHAPTER.

OF UNDEMARCATED OR DISTRICT FORESTS.

26. *Powers of Local Governments.*—For the protection of forest growth on lands which have not been formally demarcated, but which are either the property of Government, or over which Government has certain rights, the following provisions are needed. To these provisions the condition should be attached that no rights of communities or other persons will be affected or abridged by the rules containing them :—

I.—Power to protect certain species of trees, to be called reserved trees, and to prohibit or restrict the cutting, marking, lopping, using or removing of such reserved trees or timber.

II.—Power to mark or girdle trees of any kind, and to prohibit the cutting, removing, or otherwise using of such trees.

III.—Power to prohibit the cutting and removal of timber under a certain size or the fashioning or converting of any or of certain kinds of timber.

IV.—Power to prohibit the clearing of forest for cultivation, or the alienation of forest land without the permission of the Conservator of Forests or such other officer as may be appointed in that behalf by the Local Government.

V.—Power to regulate the disposal of timber and other produce of such forests by sale or otherwise, and to authorize the levying of dues on the felling, cutting, removing, or otherwise using of trees or bamboos, or on the collection or export of gums, fruits, grass, or any forest produce. (The necessity of this provision is explained in the third chapter, paragraph 34.)

VI.—Power to close certain portions of the forests, to declare all rules of demarcated public forests applicable to the blocks thus closed, and specially to prohibit all ingress of men and cattle except on authorized roads, the setting of fire to the grass and forest, and all cutting and interference with the forest without authority.

VII.—Power to notify certain forest tracts, the maintenance and protection of which may appear necessary to the Local Government for the following reasons :—

(a) for the preservation of the soil on the tops, ridges, slopes, and in the valleys of mountain ranges, and for protection against storms, landslips, and avalanches ;

(b) for the protection of the banks of mountain streams, rivers, and other waters ;

(c) for the maintenance of a water-supply in springs and streams ;

(d) for the protection of any land against shifting and moving sands ;

(e) for the protection of roads, bridges, railways, and other lines of communication ;

(*f*) for the better preservation of the public health in the vicinity of such forest.

Forests thus notified shall be called protected forests, and by such notification all cutting and burning of the forest, or of the grass in such forest, as well as all grazing of cattle and removal of produce, may be prohibited.

Local Governments should also be empowered, with the sanction of the Government of India, to promulgate rules giving effect to these provisions, under the proviso already mentioned that such rules shall not affect or abridge the rights of communities or other persons.

27. *Definition of undemarcated forests.*—These forests not being demarcated, their boundaries cannot, as a rule, be defined as precisely as those of the reserves. But obviously the area over which these provisions shall be in force must be defined in some way, nor will this be difficult. In some provinces, geographical boundaries can be laid down, and in others the definition may be similar to that used in the Central Provinces Forest Rules, which includes all waste lands which are not private property. Nevertheless, the definition of undemarcated forests will, in most cases, remain somewhat vague and undefined. Yet it is essential that Government should not relinquish all control over the forest growth in lands, which are its property, or in which it possesses forest rights. In the introduction to this report, it has been explained that the State forests in the provinces and districts here referred to now measure about 12,000 square miles, and that five years hence they may possibly be increased to 14,000. If we add to this area the forests which are under separate rules in Mysore, Berar, and the districts of Ajmere and Hazara, as well as the Government Rakhs, Trans-Indus, and those portions of the leased forests which are either demarcated, as in Chamba, or under the special care of the Forest Department, the area of demarcated forests, the protection and improvement of which has been undertaken by the Government of India, may, possibly, five years hence, amount to 16,000 square miles. This area, however, is most unequally distributed, and though it may, under good management, be sufficient to furnish the requisite material for consumption in the country and for the requirements of export trade, its influence upon the climate (if any) and upon the retention of soil and surface drainage in mountainous districts will be limited, and there may be a time when Government will come to the conclusion that the protection and improvement of the demarcated public forests has been so beneficial that it is desirable to extend their area. This extension, whenever it may take place, and whether it may be in the shape of State or Communal forests, can only be effected from the area of waste lands at the disposal of Government, and for this reason legislation must at present provide for the continuance of a certain control over the so-called open, unreserved, or district forests. The provisions suggested in the preceding paragraph have been taken from existing rules in the different provinces, they have been found necessary by actual experience, and on condition that no rights of persons or communities shall be affected, they are unobjectionable. Provisions for recording, regulating, or extinguishing adverse rights in the undemarcated forests are not required, and would not lead to any satisfactory result. The first provision—the protection of certain reserved kinds, such as teak, blackwood, sandalwood, sâl, deodar, and others—is intended to legalize the continuance of the system of royal trees which has existed for ages in many parts of India long before forest administration was ever thought of.

The 6th clause, which proposes to give power to close certain blocks, has been inserted with special reference to the system lately initiated in the unreserved forests of the Central Provinces, of closing certain blocks for a time in order to guard against the impending denudation of these lands. Whether the system will prove practicable and useful is another question; the needful legal provision, however, should be made.

The 7th clause may, in some cases, lead to the demarcation of forest tracts, but not necessarily. It is intended to cover such provisions of the rules existing in several provinces, as prohibit the clearing of belts of forest along roads, on the banks of hill streams, and around springs.

Outside the forests it may be found desirable in some cases to have the power to protect avenues, groves, or trees along public roads, canals, around public wells, buildings, burying grounds, or other public places, and the needful power may conveniently be given by the Forest law. Power to protect sacred groves, when such appears expedient for special reasons, as in Coorg, should, as already explained, be taken in the chapter on demarcated forests.

28. *Interference with private forests not contemplated.*—With private forests no interference is contemplated, beyond the general provisions for the control of timber in transit, and perhaps the power of levying duty on timber cut in these forests, which will be explained in the next chapter. Such interference could only be justified by the necessity of maintaining intact existing forests, which form natural protection belts against the encroachment of moving sands in the plains, against the drying up of springs and the action of storms, snow and water in mountainous regions. In fact, it could only be justified by considerations similar to those which are enumerated in paragraph 26 (VII) as sufficient to warrant the placing of any tract in the district forests under special protection. Whenever it may be deemed necessary artificially to create a forest for a protection belt, the land required for that purpose should, under all circumstances, be expropriated under the Land Acquisition Act; and, under existing circumstances, the maintenance of forests also, which are required to be maintained as protection belts, should be secured by acquiring them for the State. The main reason why another course is not practicable at present is, that the Forest Department in India is yet too weak and not sufficiently experienced to undertake anything beyond the management of the Government forest domains ; for it is easier to manage estates which are under our control than to exercise a practically useful control over the management of forests by other proprietors.

29. *Acquisition of land for forest purposes.*—The only provision therefore that is required at present with reference to private lands is that whenever it may appear to the Local Government desirable on public grounds to expropriate any land for forest purposes, or any forest in which Government has or has not forest rights, such expropriation shall be effected according to the provisions of the Land Acquisition Act of 1870, or other law for the time being in force, and that lands so acquired may be notified as State or village forests. The *Code Forestier* (Art. 219-226, Law of 18th June 1859) compels all proprietors of forests to report their intention to make clearances (*défrichements*), it authorizes the local Forest officers to object to such clearances, if likely to be injurious to the public welfare, and it prohibits such clearances, if within a term of six months from the date of the objection they are formally interdicted by Government. The Forest laws of several Swiss cantons have exceedingly strict provisions to regulate the management, and against the clearance or devastation of protection belts (Bannwälder). A certain control over the management of private forests is exercised in several States of Germany, and in Prussia especially the necessity of consolidating and revising the existing legislation on this important subject has lately received great attention, but, as previously explained, in India we must guard against undertaking too much in the present state of the Forest Department. Under existing circumstances, the only chance of doing justice to the task before us is by concentrating our attention as much as possible upon limited areas of demarcated public forests.

THIRD CHAPTER.

ON THE CONTROL OF TIMBER AND FOREST PRODUCE IN TRANSIT.

30. *Preliminary remarks.*—The provisions required under this head will be found in the following documents: Panjab River Rules of 1871; Bengal Rules of 1871, Part IV ; Burma Forest Rules of 1865, Chapter IV, and Mr. Baden-Powell's Draft ; Burma Timber Rules of 1873. It will also be useful to consult the Berar Rules of 1871, Part IX. The Government Forest Act of 1865 limited the control of Government to timber, the produce of Government forests, and to streams flowing from Government forests or used for the transport of the

produce of such forests. The provisions of the new law must be made applicable to all timber and forest produce, whatever its origin, and to all public rivers, streams, and waterways within certain districts or other geographical limits, to be notified by the Local Government of each province. The following provisions are essential to secure the object in view.

31. *Streams used for floating to be kept open.*—The Local Government must be empowered to prohibit the closing or obstruction for any purpose whatever of the channel or banks of any river, stream, canal, or other water which is used for the transport of timber, bamboos, or other forest produce. This provision must include power (*a*) to regulate the construction of dams for fisheries or irrigation so as not to interfere with the floating of timber; (*b*) to prohibit the throwing in of grass, branches and other material (*e. g.*, toungya refuse in the hill-streams of Pegu); (*c*) to prohibit the stoppage of, or interference with, timber while in transit, or when landed on the banks of any stream or other waterway for the purpose of, and in the course of, transit, by persons other than forest officers, or the levy of tolls or fees by such persons on such timber.

To this provision the condition should be added that no private rights are infringed or restricted. No private rights, it is true, ought to be acknowledged in public streams which are used for navigation or the floating of timber; but I am not sure whether private rights in public waters have not already in some cases been recognized in this country.

32. *Routes for the transit of timber to be prescribed.*—The Local Government must be empowered, within certain districts or territorial limits, to prescribe the routes by land or by water, by which timber, bamboos, or other forest produce, whatever their origin, shall be removed or conveyed, as well as to close any of such routes. Further to prescribe the ports or stations at which such produce shall be landed or loaded, and to prohibit the landing or loading of such produce at other places.

This provision is considered necessary in the Bombay Presidency, and the first part of it is indispensable in all Government forests, the produce of which is cut, collected, and removed by purchasers. The people who require grass, bamboos, firewood, dry timber, gums, fruits, and a variety of other forest produce, cut or collect it in the forests under certain rules, licenses or permits, and bring it out by certain roads or waterways to stations or toll-houses where it is examined, passes issued and the purchase-money or seignorage due on the produce adjusted, unless this was paid previously, when the license or permit was bought. In some cases there is a second line of stations outside, at which the produce is compelled to stop and the passes are checked. Under this system large quantities of dry timber, bamboos, and other forest produce is exported from the forests of Oudh and the North-Western Provinces and from forest districts in other provinces. In Burma it is known under the name of the permit system, which, though it has now been abolished in most forests, will probably be continued in some of the more remote districts, whence the export of timber on Government account for sale at depôts would be too expensive. It is obvious that this system cannot be maintained without power to prescribe the routes for removal, and to prohibit all routes by which the stations might be circumvented. Other provisions are needed besides, which will be found further on.

33. *Timber stations.*—The Local Government must be empowered to prescribe certain stations or places where all timber, bamboos, or other forest produce, whatever its origin, shall be reported and stopped, either for the purpose of being examined, or for the payment of any dues lawfully payable thereon, or for other purposes, and to order that such produce shall not be removed without the permission of competent authority. This provision is necessary, not only for the protection of the system of working certain Government forests as just described, but also to facilitate the protection of Government forests by controlling the timber imported from foreign territory or from private forests (Bengal, Bombay, Panjab), and for the purpose of levying dues

payable on timber imported from foreign territory or from private forests (Bengal, Bombay, Burma).

34. *Levying of duty.*—The Local Government must be empowered, under the sanction of the Government of India, to authorize the levying of duty on timber, bamboos, and other forest produce brought from any forest situated within British territory or from foreign territory, and to prescribe the rates of such duty, the manner and place of levy, and the person by whom it shall be levied. This provision will be necessary in case the new forest law is extended to the Bombay Presidency, unless, meanwhile, this matter is legalized by a local enactment. For Burma Act XIII of 1873 has given the needful power with regard to foreign timber, but it is for consideration whether it would not be convenient to have all matters relating to duty on timber in one enactment. Tolls on timber, canoes (dug-outs), grass and other forest produce are levied also on the rivers which come from the hills east of Chittagong. Under these circumstances it seems desirable to include this provision in the Forest Act, and it should be provided that Government has a lien on such produce for the duty. In connection with this subject it should be considered whether this is the proper place for the provision legalizing the levy of dues on account of seigniorage or purchase-money for timber, bamboos, and forest produce under the system described in paragraph 32. The transactions which give rise to these dues are of the character of sales, with that peculiar feature, that the purchaser selects, cuts and removes, under permit or license, or under certain rules, the articles which he intends to export, and that the transaction as between him and Government is not, unless he has made payment in advance, completed, until he has brought the produce to the duty station situated on or near the edge of the forest or outside the forest. That Government has a right of ownership in the produce is undoubted, for it is the produce of its own forests. Nevertheless, inconvenience is sure to arise unless the nature of the transaction is distinctly recognized by law. Local Governments should have the power to authorize officers in charge of Government forests to levy dues, at certain rates to be fixed from time to time for each forest by the Conservator of Forests, on timber, wood, bamboo, or other forest produce exported from these forests, or to levy (as is done in several districts of the Central Provinces) annual rates from each householder, who is permitted to provide himself with certain kinds of forest produce, or of grazing his cattle in the forest. The provisions here suggested should not be limited to demarcated forests, but should be extended to all forests under the control of Government, with the provision that the rights of individuals and communities should in no way be restricted or abridged thereby.

35. *Passes.*—The Local Government should be authorized to prescribe the issuing of passes for timber, bamboos and forest produce in transit, the form of such passes, and to prohibit the transport of such produce within certain districts, or on certain routes or rivers except it is covered by such passes. The necessity of this provision for the protection of the forests and to protect the interests of Government in the matter of duty levied on imported timber is obvious.

36. *Timber marks.*—The Local Government should be empowered to prescribe the registration of marks used as property marks for timber or other forest produce, the number of marks which may be registered for each person, the levy of fees for such registration, and the time during which such registration shall hold good. Also to prohibit the use of any marks not registered, and to prohibit except by forest officers—or other persons under special authority—on certain rivers or within certain districts the use of marking hammers or other tools or implements whose sole or chief use is for making or altering marks on timber. This rule is necessary on all rivers where timber is floated, specially on the Himalayan and Burma rivers.

37. *Unlawful conversion, &c., of timber.*—The Local Government should be authorized, within certain territorial limits, to prohibit the converting, cutting into pieces, sawing up, splitting, hollowing out, burning, chipping, concealing,

removing, or selling of all timber or certain classes of timber, without the permission of competent authority. This measure is necessary for the protection of the timber trade on rivers, and for the protection of the interests of Government when timber is removed by purchasers and paid for at stations outside the forests.

38. *Waif and drift timber.*—One of the most important subjects to be provided for in this chapter relates to the management of waif and drift wood and timber. Waif may be defined as that to which no owner can show any right or title, by reason of the impossibility of identification. All waif is *primâ facie* the property of Government, and this should be recognized in the Act. At the same time customary rights exist and are exercised, to appropriate waif thrown on to the river bank. Such rights, where established in any particular case, must of course be respected, but the right of Government to waif floating in any public stream or river must be maintained. Drift timber may be defined as timber, which from any cause is floating loose and without control, and should include all timber beached, stranded or sunk in any river or on the sea-shore, whether marked or not. At the outset it should be provided that all drift timber shall be deemed to be the property of Government, until or unless any person prove his right and title thereto. This fundamental principle recognized, Local Governments may be empowered to provide by whom waif and drift timber shall be collected, to what stations or depôts it shall be brought, and what steps shall be taken to invite claimants. Power should also be given to appoint certain persons to collect waif and drift timber, and to require all persons who have collected such timber to deliver it to the persons authorized to receive it on payment of such reward for salvage services as may from time to time be fixed by the Local Government. Power should further be given to dispose, for the benefit of Government, of timber not claimed within a certain fixed time. Authority should be given to the Conservator of Forests, or to any Forest officer authorized by him on that behalf, after due enquiry, to adjudicate the timber to any person proving his claim thereto, or in the case of several claimants to refer them to the Civil Courts, unless they consent to his arbitrating between them. Power should also be given to determine and levy certain rates of royalty and salvage to be paid by claimants to whom drift timber has been awarded, over and above the expenses which may have been incurred on account of the timber.

These are the principal provisions which find a place in Chapter III of the Bill. Local Governments should have power, subject to the sanction of the Governor General in Council, to make rules to regulate all matters arising out of these provisions.

FOURTH CHAPTER.

OF THE PREVENTION OF OFFENCES AND OF PENALTIES.

39. *Powers of Police and Forest Officers.*—Part III of the Bill framed by Mr. Baden-Powell seems to me to meet most requirements under this head, and I have only a few remarks to add and modifications to suggest.

The points which must be provided for are as follows:—

Police and Forest officers must have the duty to prevent offences against the Act, or against the rules made under it, and they must be empowered to do the following acts:—

(a) to seize and detain cattle unlawfully straying in or doing damage to any public demarcated forest; or any block closed under VI, paragraph 26.

(*b*) to seize and detain timber, bamboos, or other forest produce unlawfully removed from any forest to which the Act applies, or otherwise unlawfully dealt with, or which is not covered by a pass or other proof of ownership as prescribed by the rules made under the Act. Likewise any tools, boats, carts and cattle used in infringing any of the provisions of the Act, or of the rules made under it ;

(*c*) to arrest without warrant any person committing an offence against the Act, or the rules made pursuant to it; provided that the person so arrested is taken before a Magistrate without unnecessary delay, and that Local Governments may restrict the power of arrest without warrant to certain districts, to certain seasons, or to certain classes of offences, or only permit it under certain circumstances.

40. *Cattle trespass.*—Regarding cattle trespass the provisions of Mr. Powell's sections 64 and 65 may be adopted as they stand. The Cattle Trespass Act should be made applicable, authority being given to Local Governments to alter the scale of fines, and to provide the levy of increased fines in the case of a second offence, or if the trespass was committed between sunset and sunrise.

41. *Confiscation.*—Regarding the question under what circumstances timber, tools, boats, and other articles seized under paragraph 39 (b) are liable to confiscation, and the disposal of those articles, it will suffice to refer to Sections 71 to 74 of Mr. Baden-Powell's draft Bill. In the case of timber or other forest produce unlawfully removed from the forests, the term confiscation does not apply. When such articles are seized and retained, Government merely recovers its own property, though not in the same place and in the same condition as it was before removal. The persons who may have unlawfully cut and removed a tree, and who are forced to give it up after they have brought it outside the forests, have expended labor and money on the timber, and it might be maintained that Government gains by recovering its property at a place so much nearer the market. On the other hand, it must be considered that it may not have been the intention of Government at all to bring that tree to the market, and that great and perhaps irreparable damage may have been done to the forest in its present condition by cutting it. Under these circumstances it is evident that the produce of Government forests unlawfully removed ought to be dealt with differently from timber which was the property of the offender, and which he had taken past a duty station without paying duty, or which, for other reasons, may have become liable to confiscation. Section 68 of the draft Bill of 1869 made a distinction between the two cases, and directed that timber or forest produce should either be restored to its lawful owner, or no such owner appearing, should be confiscated and sold on account of Government. It is for consideration whether this distinction should not be maintained. Section 70 of Mr. Powell's draft relates to the same subject. This section intended, with reference to Section 418 of the Criminal Procedure Code, to empower the Court, after the trial of any criminal case is concluded, to make such order as appears right for the disposal of any property connected with the case, notwithstanding that such property, owing to its bulk, or otherwise, has not been produced before the Court. Under all circumstances Forest officers above a certain rank should be empowered, under certain forms of procedure, summarily to recover timber or other forest produce, the property of Government, which has been unlawfully removed from any Government Forest.

42. *Arrest without warrant.*—Act VII of 1865 does not limit the power of arrest without warrant, and I am of opinion that this power cannot usefully be limited in general terms, but that the limitation must be regulated by the requirements of each river or forest district, for circumstances vary extremely in this respect. When a Magistrate resides in the vicinity of a forest, or when superior Forest officers have magisterial powers, it may not be expedient to authorize subordinate forest or police officers to arrest without warrant; but in many cases the nearest Magistrate is a long way off, often 50 miles or more over difficult ground, and hardly accessible at certain seasons, and

in such cases the power of arrest without warrant is necessary. Local Governments should be empowered to define the districts in, or the circumstances under which arrests without warrant may be made.

43. *Safe-guards against the abuse of powers.*—The question of conferring any of these powers upon subordinate forest and police officers, to be exercised mostly in uninhabited places, away from any control, is not one without difficulty, and two safe-guards appear necessary, which should receive legal sanction. *First*, no officers below a certain rank, as may be prescribed by the Local Government, shall exercise any of these powers, unless they wear such uniforms or badges of office, as the Local Government may from time to time direct. *Second*, the provision of Section 12 of the Government Forest Act, providing penalties for the vexatious or unnecessary arrest or seizure must be maintained. This provision was retained in the draft Bills of 1868 and 1869, but was omitted from Mr. Baden-Powell's draft.

44. *Search warrant.*—Section 77 of Mr. Baden-Powell's draft empowers Magistrates of a district, or division of a district, on the application of a forest or police officer, to grant a search warrant for stolen or smuggled timber or other forest produce. Such cases are not likely to happen frequently, but cases have happened when search was necessary. The object of this section would, in most cases, be defeated if it was necessary to apply to the local magistrate for a search warrant. I would prefer the provisions of Sections 19 to 21 of the Inland Customs Act of 1875, and authorize Local Governments to grant the power of search to forest officers above a certain rank, and to provide that the search be made in the presence of a police officer.

45. *Persons having a beneficial interest in a forest to assist in its protection.*—Section 63 of Mr. Baden-Powell's draft and Section 58 of the draft of 1869, which requires communities or persons who have a beneficial interest in the forest to assist in the protection of the forest, is necessary, and should be adopted. With regard to protection against fires, power should be given to Local Governments to close any Forest or portion of a Forest in which a fire has occurred, against pasture cutting of wood or other user of Forest produce, whether such user were exercised in virtue of a right, a privilege or a written or verbal agreement, unless such right, privilege or agreement clearly gave or implied the right to fire the Forest or the grass growing within it. In the tropical and subtropical provinces of India where jungle fires are an annual institution, this power, if given in the Act, must perhaps for the present remain in abeyance, but for the Panjab and Sindh the provision is necessary.

46. *Penalties.*—Section 66 provides penalties, and may be adopted as framed. The principal difference between the penalties provided in this section and the corresponding provisions of the Government Forest Act is that the penalty may be either fine, or imprisonment, or confiscation, or any two of these penalties together. The cumulation of fine and imprisonment I would not press, for cases requiring such treatment will generally be dealt with under the Penal Code, but power should be given to adjudge fine or imprisonment in addition to confiscation.

As the forest rules made under this Act become more perfected, and more experience is gained regarding their working, it will be possible to specify more accurately than can at present be done, the penalty to be assigned to each offence. A beginning in this direction will be to provide that the penalties prescribed shall, as much as can be, be proportionate to the damage done by the offence, and shall be increased when the offence was committed between sunset and sunrise, or in case of resistance to lawful authority, or when the offender had previously been convicted for the same offence. All this is on the understanding, that in case of theft or fraud, or other crime or offence punishable under another law, action under such law shall not be held to be barred by the provisions of the Forest law.

Section 75, providing that portion of any fine or penalty may be paid to the informer, is obviously necessary.

47. *Power to compromise with offenders.*—A provision which will probably prove most useful may be framed analogous to the Inland Customs Act, (Section 15), to the effect that the Local Government may from time to time by rule direct, that any forest officer of and above a certain rank, if satisfied in such manner as such rule may prescribe, that any offence against the Act, or any rules made under it, has been committed, shall, instead of preferring a complaint before a Magistrate, impose such penalty but always less than the limit fixed by the Act, as may be sanctioned by the Conservator of Forests under such rules as may from time to time be prescribed by the Local Government.

This system is in force in the French State and Communal Forests; it was introduced into the *Code Forestier* (Art. 159) by a law of 18th June 1859, and its effect has been most beneficial in saving the time of forest officers and diminishing bad feeling against the forest administration. During a tour which I made through the public forests of France in 1866, I became acquainted with the great advantages of this system, but refrained from proposing any similar provision when framing the draft Forest Bills of 1868 and 1869, because I could not at the time trace any provisions of similar character in Indian legislation. The analogous provision in the Inland Customs Act, however, seems to justify my present proposal. The section of the French Forest Code stands as follows :—

L'administration des forêts est autorisée à transiger, avant jugement définitif, sur la poursuite des délits et des contraventions en matière forestière commis dans les bois soumis au régime forestier. Après jugement définitif, la transaction ne peut porter que sur les peines et réparations pécuniaires.

All " transactions" in France require the sanction of superior authority, viz., —I. In case of offences committed by purchasers of standing timber—of the Director General of Forests—when penalties, including damages, do not exceed 1,000 francs, and of the Finance Minister when they exceed that sum. II. In case of all other offences—of the Conservator when penalties, including damages, do not exceed 500 francs; of the Director General for amounts not exceeding 1,000 francs, and of the Finance Minister when they exceed that sum. The proposals are submitted by the district forest officers (*Chef de Cantonnement*).

48. *Protection of boundary marks.*—Section 78 of Mr. Baden-Powell's Bill provides for the protection of boundary marks. Some provision to this effect is probably necessary; but it seems to me preferable to prohibit (in Chapter I) any moving of, or interference with, boundary marks, and to make an infringement of that rule liable to the same penalties as other forest offences. The necessity of placing Forest boundary marks specially under the protection of the Law, is obvious. Several cases in point were reported in the Annual Report of the Northern Division of the Bombay Forests for 1873-74 (paragraphs 59, 60). In one case an Inamdar had cleared several acres of Government Forest and had attempted to conceal his proceedings by removing the boundary marks of his village so as to include in his own forest that portion of Government land which he had been plundering. The mischief done by the removal or damaging of boundary marks is in most cases much greater than the expense of restoring the marks which, with the reward to the informer, is the limit of the penalties proposed in Mr. Baden-Powell's draft.

49. *Conclusion.*—In conclusion, I desire to draw attention to certain matters which belong to the preamble of the Bill. The provisions regarding forest officers will be found in Sections 5, 6, 7 of Mr. Baden-Powell's draft. It will be convenient, for the framing of forest rules in the different provinces, at once to say that the officers subordinate to the chief forest officer in each province will ordinarily appertain to those classes whose principal functions are designated by their names, viz., controlling, executive, and protective officers. These classes of forest officers exist in all countries, and though forest administration in this country is not yet far enough advanced completely to separate those functions, they will eventually be separated, and to the officers of each class will be

assigned certain limits of powers and certain degrees of responsibility regarding forest rules.

Mr. Baden-Powell expressed doubts regarding the retention of Section 7 of his draft. It is analogous to Section 10 of the Police Law of 1861, and it is of much importance that there should be no doubt regarding forest officers being prohibited to engage or have an interest in timber trade and similar business. Its retention therefore is desirable.

I would also suggest that Section 64 of the draft Bill of 1869 be restored, which empowers Conservators to sentence any forest officer under their control, whose appointment is not notified in the *Gazette*, to pay a fine not exceeding one month's pay. It is true that this and the last-named section may be entered in a contract of service, but doubts regarding the legality of such fines or prohibitions are readily raised and entertained, and it is better to cut off all possibility of doubt in a matter of this kind.

Some definitions will be necessary at the commencement of the Act. The word "forest" does not require any definition here, but the following, which with a few modifications and additions are taken from Section 4 of Mr. Baden-Powell's draft Bill, will be found indispensable:—

"Cattle" shall, besides horned-cattle, include elephants, camels, horses, asses, mules, sheep, goats and swine.

"Timber" shall include wood, whether cut for building, fuel or other purposes, as well as bamboos, and shall also include wood fashioned or hollowed out for cartwheels, mortars, canoes or other purposes.

A mark placed upon a standing tree by a public servant, to indicate that such tree is the property of Government, or that it may lawfully be cut and removed by another person, shall be deemed to be a property mark within the meaning of the Indian Penal Code.

"Forest right" shall mean—

(a) the right of ownership in trees, timber, bamboos, brushwood, grass, or other produce of a forest;

(b) the right to cut, use, or remove any timber, trees, branches, leaves, or any bamboos, brushwood, grass, or other produce of a forest, or of the soil thereof;

(c) the right to prohibit such cutting, user, or removal;

(d) the right of pasturing cattle in a forest;

(e) the right of prohibiting such pasturing;

(f) the right of way through a forest;

(g) the right of hunting, shooting, or fishing in a forest.

"Forest privilege" means any user of a forest or the produce thereof (as included in the definition of forest right) when such is exercised by permission, and at the pleasure of the Government, and not as of right.

"Timber Station," shall mean any place which the Local Government shall from time to time appoint for the storing of wood, timber, bamboos, or other forest produce, for the stoppage and examination of wood, timber, bamboos, or other forest produce in transit, or for the detention of the same for the levy or payment of duty, royalty, or otherwise, or for the storing of drift and other timber, bamboos, and other forest produce lawfully collected under this Act, or rules made under it.

"River," includes stream, canal, creek, or other channel, natural or artificial.

Timber often is exported in the shape of canoes; so from the Chittagong Forests, various articles, such as mortars for oilseeds, pestles, basins, and the like are fashioned in the forest for export, and may conveniently be included under "timber."

The necessity of extending the definition of "property mark" is explained on page 4 of the 1874 Conference Report. If I am correctly informed, the case in Burma was that in addition to a number of trees marked by Government which could lawfully be cut and removed by the permit-holder, others were marked and girdled fraudulently, and that those who had perpetrated this fraud could not be convicted of counterfeiting a property mark used by a public servant. This will explain the modified wording of this definition. Similar cases may arise in all forests which are worked under the system adverted to in paragraph 32 of purchasers being permitted under license, permit or general rules to cut and remove timber or forest produce.

It will be a great advantage to use one term for all nakas, chaukis, posts, stations, where timber or other forest produce in transit is stopped for examination, or the payment of dues, or where drift and unclaimed timber is collected by Government. Where the transit is by water, these posts are usually called river stations. Mr. Baden-Powell adopted the term timber depôt, which however has a restricted meaning, and I would therefore suggest the term "timber station."

These remarks, together with the two revised draft Forest Bills of 1868 and 1869, with Mr. Baden-Powell's draft of 1874, and with the local rules and draft rules printed in the appendix, will make it possible to frame a brief Forest Bill, which, it is hoped, will facilitate the protection and improvement of the public forests in the Panjab, the North-West Provinces, Bengal and the provinces under the Government of India, and place other matters connected with Forest Administration upon a satisfactory footing.

APPENDIX.

I.—THE GOVERNMENT FORESTS ACT, NO. VII OF 1865.

PASSED BY THE GOVERNOR GENERAL OF INDIA IN COUNCIL.

(Received the assent of the Governor General on the 24th February 1865.)

AN ACT TO GIVE EFFECT TO RULES FOR THE MANAGEMENT AND PRESERVATION OF GOVERNMENT FORESTS.

WHEREAS it is expedient that rules having the force of law should be made from time to time for the better management and preservation of forests wherein rights are vested in Her Majesty for the purposes of the Government of India; It is enacted as follows :—

Preamble.

Interpretation Clause. 1. In this Act, unless there be something repugnant in the subject or context—

"Government Forests." "Government Forests" shall mean such land covered with trees, brush-wood or jungle, as shall be declared in accordance with the second Section of this Act to be subject to its provisions.

"Magistrate." "Magistrate" shall mean the Chief Officer charged with the executive administration of a district or place in criminal matters by whatever designation such officer is called, and shall include any person invested by the Local Government with the powers of a Magistrate or of a subordinate Magistrate as defined in the Code of Criminal Procedure, with a view to the exercise by him of such powers under this Act.

"Local Government." And in every part of British India in which this Act operates, "Local Government" denotes the persons authorized to administer executive Government in such part, and includes the Chief Commissioner of any part of British India under the immediate administration of the Governor General of India in Council whenever such Chief Commissioner is authorized by the Governor General in Council to exercise the powers of a Local Government under this Act.

Governor General in Council and the Local Governments may render certain lands subject to the provisions of this Act. 2. The Governor General of India in Council within the Provinces under his immediate administration, and the Local Governments within the Territories under their control, may, by notification in the Official Gazette, render, subject to the provisions of this Act, such land covered with trees, brush-wood, or jungle, as they may define for the purpose by such notification : Provided that such notification shall not abridge or affect any existing rights of individuals or communities.

Local Governments may make rules for management and preservation of forests, and for regulating the conduct of persons employed on them. 3. For the management and preservation of any Government forests or any part thereof in the Territories under their control, the Local Governments may, subject to the confirmation hereinafter mentioned, make rules in respect of the matters hereinafter declared, and from time to time may, subject to the like confirmation, repeal, alter, and amend the same. Such rules shall not be repugnant to any law in force.

What may be provided for by rules made in pursuance of this Act. 4. Rules made in pursuance of this Act may provide for the following matters :—

First.—The preservation of all growing trees, shrubs, and plants, within Government forests or of certain kinds only—by prohibiting the marking, girdling, felling, and lopping thereof, and all kinds of injury thereto; by prohibiting the kindling of fires so as to endanger such trees, shrubs, and plants; by prohibiting the collecting and removing of leaves, fruits, grass, wood-oil, resin, wax, honey, elephants' tusks, horns, skins and hides, stones, lime or any natural produce of such forests; by prohibiting the ingress into and the passage through such forests, except on authorized roads and paths; by prohibiting cultivation and the burning of lime and charcoal, and the grazing of cattle within such forests.

Second.—The regulation of the use of streams and canals passing through or coming from Government forests or used for the transport of timber or other the produce of such forests—by prohibiting the closing or blocking up for any purposes whatsoever of streams or canals used or required for the transport of timber or forest produce; by prohibiting the poisoning of or otherwise interfering with streams and waters in Government forests in such a manner as to render the water unfit for use; by regulating and restricting the mode by which timber

A

shall be permitted to be floated down rivers flowing through or from Government forests and removed from the same; by authorizing the stoppage of all floating timber at certain stations on such rivers within or without the limits of Government forests for the purpose of levying the dues or revenues lawfully payable thereon; by authorizing the collecting of all timber adrift on such rivers, and the disposal of the same belonging to the Government.

Third.—The safe custody of timber the produce of Government forests—by regulating the manner in which timber, being the produce of Government forests, shall be felled or converted; by prohibiting the converting or cutting into pieces or burning of any timber, or the disposal of such timber by sale or otherwise, by any person not the lawful owner of such timber, or not acting on behalf of the owner; by regulating the manner in which property-marks shall be affixed to timber and other forest produce in Government forests; by prohibiting the affixing of property-marks to timber by any person not the owner of the timber or acting on behalf of the owner, so long as such timber shall be within certain territorial limits, or shall be in transit on certain rivers; by prohibiting within certain territorial limits the effacing or alteration of property-marks on timber; by prohibiting, within such limits, the use of the property-marks employed by the Government, or the fraudulent use of the property-marks of private persons; by requiring the registry within certain territorial limits of implements for affixing property-marks on timber; by directing the levying of fees for the registration of such implements.

Fourth.—The regulation of the duties of the Government officers and establishments charged with the management and conservancy of Government forests and with the levy of forest dues and revenues—by prohibiting their engaging in any employment or office other than their duties as public servants; by fixing penalties for the wilful neglect of the rules laid down for the guidance of such persons in all matters connected with the guarding of the boundaries of the forests, the marking, girdling or felling of trees, the marking and passing of timber, the reporting and preventing of offences against the rules made in pursuance of this Act, and the collecting of forest dues or revenues.

5. In cases where the penalty of confiscation is not provided by this Act, the Local Government may prescribe punishments for the infringement of rules made in pursuance thereof, by fine not exceeding five hundred rupees, and in default of payment of such fine may provide for the imprisonment of the offender for such term as is mentioned in the sixty-seventh Section of the Indian Penal Code.

Power to Local Government to prescribe punishments for infringement of rules.

6. Such rules when confirmed by the Governor General in Council and published in the Official Gazette shall have the force of law.

Rules when confirmed and published to have the force of law.

7. All implements used in infringing any of the rules made in pursuance of this Act, and all timber or other forest produce, removed or attempted to be removed, or marked, converted, or cut up contrary to such rules, shall be confiscated.

Confiscation in case of infringement of rules.

8. Any Police Officer or person employed as an officer of Government to prevent infringement of the rules made in pursuance of this Act may arrest any person infringing any of such rules, and may seize any implements used in such infringement, and any timber liable to confiscation under this Act.

Arrest and seizure in case of infringement of rules.

9. Any person arrested on the ground that he has committed an infringement of such rules shall forthwith be taken before a Magistrate, who may, if he see reasonable cause, order such person to be detained in custody until the case shall have been disposed of.

Procedure in case of arrest.

10. Where the doing of any act is made punishable by this Act, or by any of the rules to be made in pursuance thereof, with any penalty, the causing or procuring such act to be done shall be punishable in like manner.

The causing or procuring a punishable act to be done is punishable in the same manner as the doing of the act.

11. When any timber or other property shall be seized as liable to confiscation under this Act, any Magistrate or officer empowered to enforce penalties under this Act within the district or division of a district wherein the same may be seized, may, upon information, summon the person in possession of such timber or other property, and upon his appearance, or in default thereof, may examine into the cause of the seizure of such timber or other property, and may adjudge the same to be confiscated and sold on account of the Government.

Procedure in respect of property seized as liable to confiscation.

12. Any Police Officer or officer of Government who shall vexatiously and unnecessarily seize the goods or chattels of any person under the pretence of seizing property liable to confiscation, or who shall vexatiously and unnecessarily arrest any person, or commit any other excess beyond what is required for the execution of his duty, shall be liable to a fine not exceeding five hundred rupees, or to imprisonment of either description as defined in the Indian Penal Code for a term not exceeding three months.

Penalty for vexatious seizures and arrests.

13. All fines and penalties under the rules made in pursuance of this Act shall be enforced by a Magistrate in the manner prescribed by the Code of Criminal Procedure, and the rules therein contained for the trial of cases and for appeals shall be applicable to confiscations adjudged under this Act.

Enforcement of confiscations and penalties under rules.

14. When the confiscation of any property shall be adjudged under the last preceding section, the same shall thereupon belong to and vest in Her Majesty, and a Warrant shall be issued by the Court to a Police Officer directing him to hold the property confiscated at the disposal of the Local Government.

Property on confiscation to vest in Her Majesty.

15. When any confiscation or penalty shall be adjudged under this Act, the Local Governments may, within three months after final judgment, call for the proceedings of the case, and, if they shall see cause, may direct that the seizure or any part thereof be restored, and may remit the penalty or part thereof, and direct that the offender be discharged.

Remission of penalties.

16. No suit or other proceeding shall be commenced against any person for anything done in pursuance of this Act, without giving to such person a month's previous notice in writing of the intended suit or other proceeding and of the cause thereof; nor after the expiration of three months from the accrual of the cause of suit or other proceeding.

Limitation of suits under this Act.

17. No charge of an offence under this Act shall be instituted except within six months after the commission of such offence.

Period within which charges to be brought.

18. This Act shall extend to all the Territories under the immediate administration of the Government of India and under the Governments of Bengal, the North-Western Provinces, and the Punjab; and it shall be lawful for the Governors in Council of Madras and Bombay respectively, by notification in the Official Gazette, to extend this Act to the Territories under their respective Governments.

Extent of Act.

19. This Act shall come into operation on the first day of May 1865, and may be cited as "The Government Forests Act, 1865."

Commencement of Act.   Short Title.

II.—PUNJAB.

A.—RULES FOR THE CONSERVANCY OF FORESTS AND JUNGLES IN THE HILL DISTRICTS OF THE PUNJAB TERRITORIES, PUBLISHED BY THE GOVERNMENT OF THE PUNJAB.

1. In any hill district within British jurisdiction, the Civil authorities have power to mark off any tract, plot, or ground, wheresoever situated, which they may consider to be specially adapted for the growth of timber or fuel.

2. The tract, plot, or ground so marked off may be declared to be a public preserve denoted by boundary marks, fenced and protected from trespass of all kinds. Within it the said authorities are empowered to prohibit, restrict, or regulate all felling and cutting, and to arrange for the development, preservation, and growth of the trees, shrubs, or brush-wood in such manner as may seem to them expedient.

3. The said authorities within the said limits are also competent to grant to any party privileges of cutting or felling, and to demand and receive fees from such party, and to determine the amount or rate of the fees.

4. Within the said hill districts, the Civil authorities may proclaim, publicly, prohibition, restriction, or regulation in regard to the felling or cutting, or other injurious act for any purpose whatsoever of any species of tree, shrub, or brush-wood, wheresoever it may be found growing, of whatever size or growth, whether grouped together or scattered about, which they may consider specially suited for the production of timber or fuel.

5. The said authorities, in regard to the said species of trees, shrubs, or brush-wood, may grant privileges of felling or cutting to such parties as they may think proper with or without the payment of fees, to be fixed as declared in Rule 3.

6. No person shall be entitled to object to the foregoing rules, whether relating to enclosures or to particular species of trees, shrubs, or brush-wood on the score of proprietary or manorial right, provided always that the Civil authorities do not interfere with the wood or fuel that may be really required by the occupants or owners of the land for agricultural or domestic purposes.

7. With the proviso above described, the Civil authorities within the said hill districts may prohibit, restrict, or regulate any operation which they may consider calculated to destroy or injure existing supplies of timber or fuel.

8. The setting fire to forest grass, brush-wood, or other combustible substances in a manner calculated to destroy or injure existing supplies of timber or fuel, may be absolutely forbidden.

9. The villagers, the owners, and occupants of the land, may be rendered responsible for conflagration occurring within their bounds, whether accidental or not, or by whomsoever caused, and such owner, occupant, or villager, may in this case be treated as if he had been guilty of an infraction of the rules.

10. The grazing of cattle or domestic animals of all kinds in such places or with such license as may be, in the opinion of the Civil authorities, injurious to existing supplies of timber or fuel, may be prohibited, restricted, or regulated in such manner as may be deemed ex-

pedient, provided always that the proper grounds for the grazing or pasturing of such cattle be not interfered with.

11. The owner or possessor of cattle which may be found grazing or pasturing in contravention of the foregoing rule, or in public preserves, or in other forbidden ground, may be treated as if he had been guilty of an infraction of the rules.

12. Any person who may infringe any of the above rules for the conservancy of forest and jungle may be fined at the discretion of the Civil authority to an amount not exceeding Rs. 100 for each offense; such fine may be realized by sale of personal property; and in the event of non-realization, the offender may be imprisoned for a term not exceeding three months.

13. The Civil authorities may appoint foresters, rangers, or other officers for the purpose of enforcing these rules. Such officials shall be held to be vested with due authority to this end. They may range over and examine all forests and jungles within their jurisdiction respectively; lay information of violation of the rules; bring persons before the Civil authority who may be found infringing the said rules; and execute summonses; they will also be liable to the usual pains and penalties for the abuse of their authority.

Extract from the letter of the Government of India, No. 1789, dated 21st May 1855.

7. His Excellency in Council does not *propose to disallow or even to alter* the general rules which you have submitted for the sanction of Government. His Excellency makes no objection to them as far as they go. But it is very necessary, in his opinion, that the issue of those rules to the Commissioners should be accompanied by an explanation of the reasons which have led to their being couched in terms so general, and also by directions to each Commissioner to prepare forthwith a set of rules adapted to the peculiar circumstances of his division.

* * * * * * *

B.—RIVER RULES.

Notification by the Government of India, in the Public Works Department, No. 17F., dated 11th March 1871.

THE following rules drawn up under Act VII of 1865 have been confirmed by the Viceroy and Governor General in Council, and are in accordance with Section 6 of the Act published in the *Gazette of India*:—

RULES TO REGULATE THE USE OF STREAMS AND CANALS FOR THE FLOATING OF TIMBER, THE COLLECTION OF DRIFT, UNCLAIMED AND STRANDED TIMBER, AND THE TRANSIT OF TIMBER IN THE PUNJAB.

1. THE right of floating timber by natural streams and artificial canals in British Territory is reserved in all cases, and the control of all such streams and canals as regards the floating of timber is vested in the Local Government.

2. All timber or wood found adrift in any river or water, or stranded on any shore, island, river bank, or otherwise, within British Territory, and all unmarked wood or timber, may be taken charge of by any Forest Officer, or other person specially authorized in that behalf by the Conservator of Forests, and may be brought to any Government timber depôt.

3. At the stations of Wuzeerabad, Madhopoor, Trimmu Ghât, Nadaon, Nowshera, Pulbâu, and at such other stations as the Conservator of Forests may from time to time direct, notices shall be published on the last day of every month, stating the number and description of pieces of drift and unclaimed timber brought in during the month under the provisions of Rule 2, and calling on claimants to send in the particulars of their claims. If within one month from the date of publication of such notice no claimant appear or establish his title, the unclaimed wood or timber shall be dealt with in the manner prescribed in Sections 25, 26 and 27 of the Act V of 1861 for the treatment of unclaimed property, or according to such other law as may be in force at the time regarding the disposal of unclaimed property.

4. All claims to drift timber shall be decided by the Conservator of Forests, or such officer as he may from time to time authorize in that behalf, provided that if a dispute arises between two or more parties laying claim to the same timber, the matter shall be referred to the Civil Court, unless the parties consent to the arbitration of the Forest Officer aforesaid.

5. Permission to parties wishing to collect their own timber which may have gone adrift, will be granted on application at the discretion of the Forest Officer in charge of the river. The permit must show the marks which should be on the timber, and the holders will only be entitled to take away such timber as bears those marks.

6. Persons who have saved any timber shall not be entitled to retain the same after tender of the authorized payment (the rates will be published in the *Gazette*) for salvage service; but shall on tender, as aforesaid, deliver the same to any person having authority from the officer in charge of the river to receive it.

7. No person whose timber may have been brought to any station as drift or unclaimed timber, shall be entitled to remove the same before payment of the salvage or other expenses (if any) incurred in bringing it in. If the salvage or other expenses are not paid, and the timber redeemed within one month from the date of issuing a written notice by the Forest

Officer, and calling on the owner to redeem the same, the Forest Officer shall be entitled to sell the wood, and after deducting the sum due for salvage or other expenses for the benefit of Government, pay any balance that may remain to the owner of the timber sold.

8. No person not being the lawful owner, or acting on behalf of the lawful owner, shall, without the permission in writing of the Conservator of Forests, or other Forest Officer specially appointed on that behalf by the said Conservator of Forests, or by decree or order of a competent Court, convert, cut up, burn, remove, conceal, sell, or otherwise dispose of any wood or timber while in transit or stranded on the banks of any river, or lying in any other place whatsoever.

9. No person shall remove or efface any property-mark from any timber while in transit on any river or water, or stranded on the islands or shores thereof, nor shall he put any mark upon any timber which is not his own, nor shall he use any Government mark, except with the permission of the Conservator of Forests, or the Forest Officer in charge of the river.

10. Any person who shall offend against the provisions of rules 8 and 9 shall be liable to a fine not exceeding Rs. 200, and in default of payment of such fine, to simple imprisonment for a term not exceeding three months.

C.—RAWAL PINDI FOREST RULES.

NOTIFICATION.

THE tracts of land in the district of Rawal Pindi hereinafter specified, are hereby declared to be Government forests for the purposes of Act VII of 1865.

I.—All the forests in the Murree and Kahuta Tahsils hitherto known as "first class rakhs."

II.—All jungle lands known as "second class rakhs" demarcated at settlement, in the Tahsils of Rawal Pindi, Fatteh Jang, Pindi Gheb, Attock, and Gujar Khan.

This notification shall not be deemed to abridge or affect any existing rights of individuals or communities in the said tracts.

RULES.

WITH the sanction of the Governor-General in Council, the following rules are prescribed under Section 8 of Act VII of 1865 for the management of the Government forests defined in Punjab Government Notification No.   dated

Explanation.—Nothing contained in these rules shall in anywise abridge or affect any existing rights of individuals or communities in respect of the lands to which the rules relate.

SECTION I.—*Of the Murree and Kahuta Forests, known as first class rakhs.*

I.—The officer of the Forest Department authorized in that behalf by the Conservator, shall select portions of the forest area, not exceeding in the aggregate 30 per cent. of the whole, and shall demarcate the selected portions by pillars or otherwise as he shall deem necessary.

The portions so selected and demarcated shall thereupon be closed absolutely against all forest rights or privileges, and shall be called "reserved forests."

Provided that, if by the reservation of any tract, any community or individual, though not having any legal right, be in the judgment of the Conservator of Forests put to special loss or inconvenience, it shall be competent to the Conservator to make suitable provision for the exercise of grazing and for the supply of fuel and timber (for domestic and agricultural purposes only), either in the reserved tract or in some adjacent tract conveniently situated.

II.—The remaining portions of forest area, not being less than 70 per cent. of the whole, shall be called "unreserved forests," and shall be open to all existing village communities, as heretofore, for the exercise, free of charge, of the following privileges only :—

(a.) Grazing or cutting grass for their own cattle.
(b.) Cutting fuel for their own use.
(c.) Cutting timber or wood for their own domestic and agricultural purposes.

III.—In unreserved forests, land on which trees stand or a growth of young trees exists shall not be cleared for cultivation or for any other purpose, except with the permission, in writing, of a Forest Officer duly authorized to grant the same.

Explanation.—Such permission shall not be requisite for the clearance in order to cultivation of land free from trees.

IV.—In unreserved forests, no person whatsoever shall be entitled to cut for sale, or to sell fuel or timber, or to burn charcoal, lime, or surkhi kilns, except upon terms of paying the authorized dues to the Forest Officer on behalf of Government.

V.—Persons not entitled to the privileges described in Rule II shall not graze their cattle in unreserved forests, except upon terms of paying the authorized dues to the Forest Officer on behalf of Government.

VI.—Forest Officers duly authorized in this behalf may grant written permits to any persons whatsoever to cut timber or fuel, and to graze cattle or to cut grass in the unreserved forests, and may receive payment for the same.

Provided that every such permit shall be deemed subject to the restrictions hereinafter prescribed, and that the Forest Officer granting the same shall take due care that the supply requisite for the wants of any village community interested be not endangered or curtailed.

Provided further, that no considerable permit shall be given without previous local inspection by a European Forest Officer.

VII.—In any unreserved forest, whenever the Forest Officer desires to protect a grove or plantation of young trees from the ingress of cattle, he shall cause an efficient fence to be erected for the purpose, and shall keep it in repair at the expense of the Forest Department.

VIII.—On the main road between Rawal Pindi and Murree, *bond fide* travellers may graze their animals, cut grass, and get dry wood for fuel on the west (or left) side of the road free of charge, but not on the east (or right) side.

Baniahs and bazar inhabitants, residing on the line of road, shall have similar privileges but subject, in regard to grazing cattle, to payment of the authorized dues to the Forest Officer on behalf of Government.

SECTION II.—*Of the fuel tracts in the Tahsils of Rawal Pindi, Fateh Jang, Pindi Gheb, Attock, and Gujar Khan.*

IX.—The fuel tracts known as second class rakhs having been demarcated and set apart at the settlement, shall be deemed to be "reserved forests," and shall be worked and managed accordingly.

X.—To prevent inconvenience to villagers and others who have been accustomed to look to such forests for the support of their cattle, on terms of due payment for grazing being made, it is hereby ordered that hereafter, whenever the Forest Officers shall have determined to close any portion of such forest against grazing, six months' notice shall be given, through the civil officers of the district, previous to such closing. The closing shall commence and the six months be calculated from the 1st day of March in each year.

SECTION III.—*Of the working of Reserved Forests.*

XI.—The supply of material from all reserved forests will be conducted according to directions laid down in the working plans, or according to the instructions issued from time to time by the Conservator of Forests, as the case may be.

XII.—The published list of seigniorage rates will not apply to trees in reserved tracts of especially fine growth or valuable quality, but such trees will be sold, when proper to do so, at special contract rates.

XIII.—All trespass by human beings or cattle, and ingress into the reserved forests except on authorized roads and path-ways is prohibited.

SECTION IV.—*Of the working of the Unreserved Forests.*

XIV.—All applications for permits shall be made in writing.

XV.—All permits granted by Forest Officers under these rules should be drawn up in such general forms as the Conservator of Forests may, from time to time, prescribe, and shall specify the date up to which they shall continue in force, and the special conditions, if any, upon which they are granted.

All forest produce, cut in contravention of the conditions of a permit, and all such produce not removed within the time specified for the permit to continue in force, shall be forfeited to the Government.

Any permit may be cancelled by the officer granting the same, upon breach of any of the conditions of such permit.

XVI.—Applications for permits to fell trees bearing a price in the Forest Department list, shall specify the brand or mark which the applicant proposes to use.

XVII.—The holder of such a permit, in carrying out the same, shall affix his brand or mark on a cleared space on the stump, within 12 inches of the ground.

Exception.—This rule shall not apply to cases where a permit shall have been given to clear a definite area of stunted trees for fuel and the like objects.

XVIII.—Permits to fell trees shall not be granted except in respect of particular localities (to be specified by the Conservator), and shall apply only to trees previously blazed or marked with the Government mark. Every such permit shall define the local area over which it is to extend.

XIX.—No trees shall be felled within 200 feet of a public road, or in the vicinity of any village site, tomb, well, spring or stream, "ziarat" or sacred place, except with the written permission of a Forest Officer duly authorized to grant the same.

XX.—Permits to collect dry fuel or dead wood for fuel will be granted as heretofore as regards small quantities, without restriction to locality. Permits to collect large quantities of dead wood will not be granted.

XXI.—Applications for permits to graze cattle or to cut grass shall specify the general locality within which, and the number and description of cattle in respect of which, the permit is desired.

SECTION V.—*General.*

XXII.—No person shall set fire to grass, trees, brush-wood, or stumps in the vicinity of any forest, reserved or unreserved. Travellers will be held responsible for the extinction of their camp fires, and persons kindling kilns of any kind will be responsible to prevent any conflagration or damage to the forest land which may occur therefrom.

XXIII.—All grazing, cutting grass, felling, lopping, maiming or injuring trees by other than right-holders or permit-holders, is prohibited.

XXIV.—Permit-holders and right-holders are prohibited from causing unnecessary damage in felling and removing materials.

SECTION VI.—*Penalties.*

XXV.—For every act or omission constituting an infringement of these rules, the offender and any one abetting his offence (within the meaning of the Penal Code), shall be liable to a fine not exceeding Rs. 500, and in default, to such imprisonment as is provided in Section 67 of the Indian Penal Code.

D.—HAZARA FOREST REGULATION.

NOTIFICATION—By the Government of India in the Foreign Department, No. 35 R., dated the 25th February 1873.

THE following Regulation for the conservancy of trees and forest lands in Hazara is published for information :—

Preamble. Whereas by the forest rules sanctioned by the Government of India on the 1st May 1855, and by the supplementary rules for Hazara, sanctioned by the Chief Commissioner of the Punjab on the 19th January 1857, Government assumed to itself the right to conserve all trees and forest lands in the said district.

And whereas, consequent on the increased demand for timber and fuel, it is necessary to make better arrangements for the said conservancy, and to define more exactly the matters in which that conservancy consists, the following Regulation for the conservancy of trees and forest lands in the Hazara district having been proposed by the Lieutenant-Governor of the Punjab, and having been taken into consideration and approved by the Governor General of India in Council, and having received the Governor General's assent, is now published with reference to 33 Vic., cap. 3, sec. 1.

PRELIMINARY.

Definitions. Forest land. 1. In this Regulation *forests* or *forest land* includes all uncultivated hill land, except public ways, grave-yards, sacred places, banks and corners of fields, habited sites, and the land immediately attached to such sites.

Uncultivated. No land shall be considered "uncultivated" within the meaning of the preceding clause which may be entered as "*cultivated*" or "*fallow*" in the faired records of the settlement now being made.

Zemindars. The word *zemindars* means persons who have a prescriptive right to the user of forest land; but this Regulation shall in no case be construed, so as to give any person a greater right of user than he possesses independently of this Regulation.

Hazara. This Regulation shall not apply to the hereditary territory of the Nawab of Umb.

So much of the Regulation published in Notification No. 31P., dated 5th January 1872, as relates to the conservancy of trees and forest lands in Hazara, is hereby repealed.

DIVISION OF THE FOREST LANDS INTO RESERVED AND UNRESERVED TRACTS.

Reserved and unreserved tracts defined. 2. At the present settlement, the forest land shall be divided by the Settlement Officer into reserved and unreserved tracts. The reserved tracts shall include all valuable forests and all forest land, the close conservancy of which is called for in the public interests, and can be effected without unduly restricting the necessary usances of the agricultural population. All forest land not included in the reserved tracts is referred to in this Regulation as the unreserved tracts.

Exceptional provisions in reserved tracts. 3. If in any case it may be found impossible to carry out this division without excluding from the "*reserved*" tract a valuable and important forest, the Settlement Officer may reserve the forest, subject to such exceptions from the sections of this Regulation applicable to the reserved tracts as may be necessary to secure to the zemindars the due supply of their domestic, agricultural, and grazing wants and other necessary usances: the said exceptions to be recorded in a settlement proceeding, and in all other respects such tracts are to be treated as "*reserved.*"

Also necessary rights of way through reserved tracts, or to drinking places therein, shall be maintained to the zemindars: but no rights of this description shall be claimable unless they are recorded in the settlement demarcation proceedings.

Boundaries of reserved tracts. 4. The limits of the reserved tracts shall be indicated by conspicuous boundaries. The boundaries to be enacted by the Settlement Officer, and thereafter to be maintained by the Forest Department.

CONTROL AND MANAGEMENT.

5. Both the reserved and the unreserved tracts will be alike under the control of the Forest Department, as regards all matters of forest administration and forest revenue dealt with by this Regulation, except so far as is hereinafter otherwise expressly provided.

Control.

A.—UNRESERVED TRACTS.

Prohibited acts in unreserved tracts.

6. In these tracts the following acts are prohibited :—

Breaking up land for cultivation, except after consent of Deputy Commissioner obtained as provided in Section 8.

Setting fire to grass tracts, or negligently permitting fire to extend thereto.

Setting fire to brush-wood, trees, or stumps of trees.

Girdling, lopping, barking, boring for turpentine, or otherwise injuring growing trees.

Removing soil or dead leaves from under trees.

Felling standing trees of the descriptions included in the seigniorage list [see Section 12], or burning kilns, without a valid order under Section 16 or 17.

Cutting young trees under any circumstances.

Felling for fuel trees included in the seigniorage list [see Section 12].

Rights of the zemindars in the unreserved tracts defined.

7. The zemindars' rights in the unreserved tracts are maintained to the following extent and no further :—

(1). They shall not use the waste products for the purpose of lime, surkhi, or charcoal burning.

(2). They may, within reasonable limits, apply to their own uses, domestic and agricultural, any tree included in the seigniorage list [Section 12] with the permission previously obtained of the Deputy Commissioner or of such other officer as the Deputy Commissioner may appoint for this purpose. No seigniorage fees shall be charged in respect of trees so used.

(3). Subject to the payment to the zemindars of half the seigniorage fees as provided in Section 13, and subject to the claims of the zemindars provided for in the preceding clause, the trees included in the seigniorage list are the property of the State. And the right to sow, plant out, and reproduce trees included in the seigniorage list is reserved to the State: provided that no land be fenced pursuant to this reservation, except for the purpose of protecting young trees; and such fencing shall not be maintained longer than is required for the safety of the young trees.

(4). Land shall not be brought under cultivation except as provided in Section 8.

(5). Subject to the exceptions and restrictions aforesaid, the unreserved tracts are the property of the zemindars; and no forest fees shall be charged to the zemindars on account of the use and enjoyment thereof, except so far as is warranted by the foregoing reservations.

8. If the zemindars desire to bring under cultivation any portion of the unreserved tracts, they shall first apply to the Deputy Commissioner for permission to do so. If the land mentioned in the application is covered with valuable forest trees, permission to cultivate it shall ordinarily be refused. Neither shall permission be granted, if the contemplated cultivation would on any other grounds be injurious to forest conservancy. In the absence of such objections, an order shall be passed permitting the cultivation of the land applied for, and the said land shall cease to be forest land from the date of the said order.

Procedure for bringing under cultivation land in the unreserved tracts.

If the zemindars fail to bring the land under cultivation pursuant to this permission within three years from the date of the order granting it, the permission shall lapse, and the land will revert to the forests. *Bond fide* cultivation of trees for marketable purposes is cultivation within the meaning of this section. Notice of all permissions to cultivate, granted or revoked under this section, shall be sent to the local Forest Officer.

9. Saving such rights as are maintained to the zemindars by Sections 7 and 8, and excepting the especial prohibitions prescribed in Section 10, the unreserved tracts shall be subject to the same provisions as are prescribed in this Regulation for the reserved tracts.

Unreserved tracts subject to the same provisions as the reserved tracts, with stated exceptions.

B.—OF RESERVED TRACTS.

10. All the prohibitions, except the first detailed in Section 6, for unreserved tracts, will also apply to reserved tracts *mutatis mutandis.* In addition thereto, the following acts are prohibited in reserved tracts :—

Prohibited acts in reserved tracts.

Grazing or driving cattle or flocks.

Cutting grass or brush-wood, or collecting fodder.

Collecting or selling fallen timber.

Collecting gums, resins, honey, wax, or other minor forest produce.

Carrying or kindling fire.

Carrying any implement to cut wood, except it is carried in pursuance of a permit to cut.

Cultivating land or preparing it for cultivation.

Squatting or building.

And it is to be distinctly understood that the zemindars may not exercise in the reserved tracts the rights maintained to them in unreserved tracts by Section 7, except where and so far as special provision to that effect may be made by the Settlement Officer under Section 3.

11. Subject only to the rights maintained under Section 3, and to the payment to the zemindars of their share of the seigniorage dues, as prescribed in Section 13, the reserved tracts shall be the property of the State. And the whole of the forest income accruing from them, other than the said seigniorage share due to the zemindars, shall be credited to the State as forest revenue.

Reserved tracts to be the property of the State.

FELLINGS, SEIGNIORAGE, AND FOREST INCOME.

12. The existing list of seigniorage fees for trees felled and kilns burnt is hereby maintained, and it shall be open to revision by the Local Government after the lapse of five years from the date of its promulgation. Subsequent schedules shall be similarly revised from time to time, but not oftener than once in every five years, provided that trees may be at any time added to, or excluded from, the list.

Assessment of seigniorage.

The seigniorage list may include all valuable timber trees, and the fees denoted in it shall be levied in respect of all such trees felled or kilns burnt on forest land other than fellings exempted by Clause 2 of Section 7.

The fees may be uniform throughout the district or vary for different tracts.

13. Half of the seigniorage fees levied under Section 12 shall be credited to forest revenue, the other half shall be paid to the zemindars within whose bounds the trees are felled.

Zemindar's share of the seigniorage.

In cases in which the Forest Department may fell trees or burn kilns on its own account or in which Government may direct that its own share of the seigniorage fees shall not be levied, the share due to the zemindars shall be paid notwithstanding.

The Forest Department shall from time to time remit to the Deputy Commissioner the sums due to the zemindars under this rule, specifying in each remittance the amount due to each village. And the sums thus received by the Deputy Commissioner shall be distributed by him to the zemindars entitled to receive them once a year as soon after the 1st day of January, as is practicable.

To enable the Deputy Commissioner to check the correctness of the remittances made duplicates of all felling orders and of orders to burn kilns shall be sent to the Deputy Commissioner by the officer issuing them the same day as they are issued.

14. The tender of the prescribed seigniorage fees by any person shall not entitle him as of right to an order for felling the trees or burning the kilns covered by the fees, unless the fees be accepted by an officer competent to issue such orders. But the fellings and kiln-burnings shall be regulated by the Forest Department subject to the general instructions of the Local Government.

Regulation of fellings, &c.

15. Seigniorage fees shall in every case be paid in full in advance prior to the issue of any order to fell trees or burn kilns.

Seigniorage fees shall be paid in advance.

16. Every order to fell trees or burn kilns shall be in writing.

Felling orders and their conditions.

Such orders shall state the amount of fees paid, and the village within the bounds of which the order shall be executed. They shall not be transferable, nor may they be executed in any other village than that originally named in the order.

Such orders shall be subject to such rules as regards cancellation and refund of fees paid as regards the selection of the timber to be felled, or the site of the kilns to be burnt; as regards the manner and time of felling the timber or of burning the kilns; and as regards the time within which the timber or the outturn of the kilns shall be removed from the forest lands, as the Conservator of Forests may from time to time prescribe with the sanction of the Local Government. For any breach of such conditions the felling order or kiln order, in respect of which the breach has occurred, may be revoked by the Forest Officer who issued it, and the timber kiln and kiln-burnings, acquired pursuant to the order, which are then still within forest bounds, may be forfeited; nor shall any claim lie for the refund of the fees paid in respect of the revoked order: provided always that a printed copy of the said conditions in the Urdu language was annexed to the order at the time of its issue.

17. Felling orders issued under Clause 2 of Section 7 shall not be subject to the provisions of the preceding section. But they shall be in writing, and a duplicate of each shall be sent at the time of issue to the executive Forest Officer. Such orders shall lapse if the trees covered by them are not felled within three months from their date.

Felling orders issued to zemindars, Section 7, Clause 2.

OFFENCES.

18. Any person who commits, or [within the meaning of the Indian Penal Code] abets the commission of any of the acts prohibited by Sections 6, 10, and 26 in the tracts respectively concerned, or who voluntarily neglects the duties imposed on him by Section 21, shall, on conviction before a Magistrate, be punished with fine that may amount to Rs. 100, or to ten times the value

Breach of this Regulation.

G

of the forest produce injured or illicitly taken, and in default of payment with imprisonment of either description for a term not exceeding six months. In the case of a second conviction, rigorous imprisonment not exceeding six months may be awarded instead of fine : provided that no sentence of fine or imprisonment shall be awarded by any Magistrate in excess of that which he is competent to award in the exercise of the powers with which he has been invested under the Criminal Procedure Code.

19. The penalty contained in the preceding section may, in the discretion of the adjudi-

Forfeiture of implements, &c.

cating Magistrate, be accompanied by the forfeiture of all implements, cattle, or conveyances used in the commission or furtherance of the offence adjudicated, and by the forfeiture of all wood or other forest produce obtained thereby.

20. When the offence found is the illicit cultivation of forest land, the forfeiture authorized

Treatment of illicit cultivation.

in the preceding section may extend to all the produce [growing or reaped] found on the said cultivation, as well as any huts or sheds erected on the land; and the sentence may further direct that the cultivator be ousted forthwith.

21. It shall be the duty of all zemindars to aid in

Responsibilities of zemindars.

the extinguishing of forest fires and in the prevention of forest offences in their vicinity.

22. In cases of serious conflagration occurring in the forest lands, or in any case in which

Communities may be proceeded against.

it may appear that any community of zemindars neglects to render reasonable assistance to the Forest Officers in the prevention and prosecution of forest offences, it shall be lawful for the Deputy Commissioner to treat the zemindars in whose vicinity the conflagration has occurred, or the community so defaulting, jointly or severally, as themselves guilty of the said acts, and to sentence them to the fine prescribed in Section 18, and in default of payment to imprisonment as therein laid down. Every such conviction may, in the discretion of the Deputy Commissioner, carry with it the deprivation of all forest dues at the time outstanding to the credit of the convicted zemindars under Section 13.

23. All cattle found straying or unlawfully grazing in any reserved forest may be seized

Cattle trespasses in reserved tracts.

by any forest official, and when so seized shall be driven forthwith to the nearest pound, to be there dealt with in the manner provided by law for cattle impounded for trespassing on cultivated land.

But where cultivation now recorded in the settlement papers closely adjoins the boundary of a reserved tract, no cattle straying from such cultivation or its neighbourhood into the adjoining reserved forest shall be seized under this section, unless the boundary line adjoining the cultivation has been efficiently fenced by the Forest Department.

And where a right of way through a reserved tract is maintained, it shall be held to protect from seizure under this section all cattle lawfully driven along the road or track indicated, even though they stray off into the forest, so long as the driver uses reasonable diligence to prevent the cattle from straying, and to drive back such as stray.

24. Convictions under Section 22 shall be appealable to the Commissioner; all other

Appeals.

criminal proceedings under this Regulation shall be appealable or otherwise in the manner provided in the Criminal Procedure Code.

MISCELLANEOUS.

25. In any case in which not less than two-thirds of the zemindars having rights in the

Voluntary conservancy.

particular piece of land in question may so desire, it shall be lawful for the Settlement Officer or for the Deputy Commissioner, after the conclusion of the settlement operations, to set aside out of the unreserved tracts a given portion for strict conservancy, and to record such rules in connection therewith as the zemindars may desire and are not objectionable on general grounds, the arrangement to hold good for the period of settlement, or for such shorter period as not less than two-thirds of the zemindars may desire. Breaches of such rules shall be punished in the manner provided under Sections 18 to 24.

Injury to trees generally prohibited.

26. The following acts are prohibited irrespective of their connection [or otherwise] with forest lands :—

Injuring or allowing cattle or flocks to injure any trees, groves, or gardens on the sides of roads in cantonments or elsewhere: provided that the said acts be not committed in exercise of a private right.

Cutting or injuring trees, which fringe or overshadow natural streams or springs.

Cutting or injuring trees, or brush-wood, or gardens situate in any grave-yard, ziárát, or other sacred place.

27. The Settlement Officer shall cause to be prepared for the Forest Department such

Copies of maps, &c., to be made for Forest Department.

vernacular copies of his maps and records as shall be necessary to enable that Department to give effect to this Regulation, without the necessity of constant reference to the settlement records; and when any subsequent alterations are made in the area of forest lands, similar documents shall be supplied to the Forest Department. The cost of preparing these records shall be paid by the Forest Department.

28. Such provisions of this Regulation as are not already in force shall come into force
from the 1st day of April 1873. As regards forest land
reserved subsequently to that date, the provisions relating
to reserved tracts shall have effect therein from the date of the Settlement Officer's order
reserving them.

Date of operation of this Regulation.

29. This Regulation shall cease to be in force on and after the 1st day of April 1874.

NOTIFICATION—By the Government of India, in the Foreign Department, No. 301R., dated 22nd December 1874.

THE following Regulation continuing for a time the Hazara Forest Regulation of February 1873, which expired on the 1st April 1874, is published for general information :—

Whereas the Regulation for the conservancy of trees and forest lands in Hazara, published under the notification by the Government of India, in the Foreign Department, No. 35R., dated the 25th February 1873, has ceased to be in force, and it is expedient to revive and continue the operation of the said Regulation for a time.

A draft of the following Regulation, together with the reasons for proposing the same, having been proposed by the Lieutenant-Governor of the Punjab to the Governor General in Council, and having been taken into consideration, and approved by the Governor General in Council, and having received the Governor General's assent, is now published with reference to 33 Victoria, Chapter 3, Section 1.

The Regulation for the conservancy of trees and forest lands in Hazara, published under the notification by the Government of India, in the Foreign Department, No. 35R., dated the 25th February 1873, shall be deemed to be and to have been in force in Hazara district from the 1st day of April 1874, and shall continue in force in the said district until expressly repealed.

III.—NORTH-WESTERN PROVINCES.

DRAFT RULES FOR THE MANAGEMENT AND PRESERVATION OF GOVERNMENT FORESTS.

THE following rules for the management and preservation of the Government forests in the North-Western Provinces, made under Section 3 of Act VII of 1865, have been confirmed by His Excellency the Governor General in Council, and are hereby published for general information :—

I.—Nothing in the following rules shall be held to abridge or affect any existing rights of individuals or communities in Government forests.

II.—Every Government forest, when not sufficiently demarcated by roads, streams, or other existing natural or artificial boundary marks, shall be demarcated with such marks as the Conservator of Forests may direct.

III.—The Conservator of Forests may, with the previous sanction of the Collector or other Chief Revenue Officer of the district, close any road or pathway traversing any Government forest. Provided always that no legal public right of way is infringed by such closing, and that no legal private right of way is abridged or affected, except after due compensation and settlement according to the law for the time being in force regarding the expropriation of such rights.

IV.—The following acts are prohibited in Government forests :—
1. Voluntarily or negligently setting fire to the forests.
2. Felling, girdling, lopping, burning, barking, stripping off leaves, tapping for resin, or otherwise injuring any trees, shrubs, or bamboos, otherwise than with lawful permission under these rules.
3. Cultivation without the permission in writing of the Conservator of Forests.
4. Grazing or pasturing of cattle, elephants, or pigs, except with lawful permission under these rules.
5. Cutting grass, collecting fruits, honey, wax, bark, or any kind of forest produce, without lawful permission under these rules.

V.—Whenever any legal rights to timber, grazing, or other forest produce, exist in any Government forest, such rights shall in all cases be ascertained and fully recorded by the revenue authorities of districts, and a copy of such record shall be deposited for reference in the forest office of the division in which the forest is situate.

VI.—It shall be lawful for the Local Government to close any portions of the Government forests, and to notify such portions as closed forests. The boundaries of such closed forests shall be demarcated by such marks as the Government may direct.

VII.—In closed forests all trespass by men, cattle or elephants off the authorized roads and pathways, is prohibited.

VIII.—It shall be lawful for the Local Government to grant in any Government forest such privileges as may be consistent with the due maintenance of the forest. *Provided always* that the exercise of any privilege under this rule shall be for the use of the person entitled thereto, and not for the purposes of export or merchandise.

IX.—No privileges granted under the last preceding rule shall be exercised in any closed forest.

X.—All timber found adrift within such limits and on such rivers as may from time to time be notified by the Local Government may be collected and taken charge of by any Forest Officer.

The collection of such timber within these limits by any person other than a Forest Officer is prohibited. All timber collected by the Forest Officer as aforesaid shall be brought to or stored at such stations as the Conservator of Forests shall from time to time publicly notify, and shall be dealt with as provided in the following rules.

XI.—Notices shall be published every second month at the three tahsil or police offices nearest the place of storing, stating the number and description of pieces of drift and unclaimed timber brought in under Rule X during the two months preceding.

XII.—All claims to drift timber shall be decided by the Conservator of Forests, or such other Forest Officer as he may authorize in that behalf: Provided that if two or more persons are claimants to the same timber, the Conservator or other officer as aforesaid may decline to decide the case, and refer the parties to the civil court.

Timber, wood, or bamboos awarded to claimants must be redeemed by the payment of all expenses incurred on the collection, landing, and protection thereof.

If the expenses are not paid and the timber redeemed within one month from the date of issue of a written notice by the Forest Officer calling on the owner to redeem the same, the Forest Officer shall be entitled to sell the wood; and after deducting the sum due for such expenses, shall pay any balance that may remain to the owner of the timber sold.

XIII.—For every breach of these rules for which the penalty of confiscation is not provided by the Act, the offender shall be liable, on conviction before the Magistrate having jurisdiction in the case, to fine not exceeding Rs. 500, or, in default of payment, to such imprisonment as is provided in Section 67 of the Indian Penal Code.

IV.—OUDH.

NOTIFICATION—By the Government of India, in the Public Works Department, No. 27 F., dated 24th September 1866.

THE following rules for the better management and preservation of the Government forests in Oudh, drawn up under Act VII of 1865, have been confirmed by the Viceroy and Governor General in Council, and are, in accordance with Section 6 of the Act, published in the *Gazette of India:—*

FOREST RULES—OUDH.

The following rules are published for the administration of the Government forests in the province of Oudh:—

I.—The officers appointed for the administration of the forests are—

1.—The Conservator of Forests.
2.—The Assistant Conservators.
3.—The subordinate Forest Officers, both in the forests and at timber stations.

II.—The administration of the Government forests throughout the province of Oudh shall be vested in the Conservator of Forests.

III.—There will be a subordinate Forest Officer in charge of every forest division. He must reside within or close to the forests. He will be expected to be thoroughly acquainted with every part of the forest, and whatever happens therein. In cases of unauthorized felling, or other breach of the forest rules, it will be his duty to report the occurrence immediately. It will further be his duty to prevent, by all means in his power, the continuance or repetition of acts constituting a breach of these rules. He will seize all wood or other forest produce unlawfully cut or removed, which he may find in the forests, or in transit from the forests; should any wood or other forest produce unlawfully cut or removed from the forest be found beyond the limits of the province of Oudh, the case shall be reported for the information of the Chief Commissioner of Oudh, and for his orders thereon. The subordinate Forest Officer will use every lawful means for the defence of the property entrusted to his charge. He will be held responsible that no trees of the reserved woods, *viz.*, Toon, Sissoo, Sâl, Ebony, Dhao, and Assana, also Khair and Tikooe or Huldoo (added by notification of the Government of India, in the Public Works Department, No. 15 F., dated the 20th August 1868), except those selected by the Conservator or his Assistants, be felled, and that no other wood be cut without the permission of the Conservator.

IV.—No Forest Officer shall engage in any employment or office whatever other than his duties under these rules, unless expressly permitted to do so by the Chief Commissioner of Oudh.

V.—The boundaries of the Government forests will be marked off as the Chief Commissioner may direct. Where the boundary line is not an established road, the bed of a river, or other line easily traced, the marks must not be more than 200 yards apart. The officer in charge of a forest division must, twice a year, at the commencement and the close of the working season, go along the whole of his boundary lines, and report the state of the boundary lines and marks to the Conservator.

VI.—Within the boundaries of a Government forest no trees of any kind shall be felled, cut, mutilated or lopped, without the special permission of the Conservator, his Assistant or the subordinate officer in charge of a division.

VII.—Permission will be granted to all villagers, living in the forests or within three miles of the Government boundary line, to cut such timber as they may require for their own *bonâfide* use, for domestic and farming purposes—such timber not being any of the eight reserved woods. (See notification by the Government of India, in the Public Works Department, No. 15F., dated 20th August 1868.)

VIII.—For purposes of merchandise, timber of any kind can only be obtained by a requisition, sent through the officer in charge for the sanction of the Conservator, or Assistant Conservator, and on payment of the specified dues.

IX.—Any person or persons detected in mutilating or causing the mutilation or injury of any kind of timber, the property of Government, will, on conviction, be liable to punishment under these rules.

X.—Any person or persons marking, girdling or felling trees, without the sanction of the Conservator, will, on conviction, be liable to punishment under these rules.

XI.—Any person or persons removing from the Government forests any timber or forest produce, without the written authority of the Conservator, will, on conviction, be liable to punishment under these rules.

XII.—Whoever, within the Government forests, sets fire to the grass or forest, or clears the jungle for the purposes of cultivation, or burns lime or charcoal, or puts cattle to graze, will be liable to punishment.

XIII.—All persons burning the grass beyond the limits of the Government forests, must take precautions that the fire be not communicated to the grass and jungle within the Government reserved forests. Should Government property be injured or destroyed by the carelessness of those burning the grass outside the forests, the persons by whose negligence the loss to Government has occurred will, under these rules, be liable to punishment.

Travellers are required to use due caution in the extinction of their camp fires.

XIV.—The use of existing Government roads through the Government forests shall be free, if compatible with the conservancy of the forests. But the Conservator, with the sanction of the Chief Commissioner, shall be at liberty to close any such roads through the Government forests, should he deem it requisite to do so.

XV.—The disposal of timber from the Oudh forests, either by sale or gratuitously, will take place according to the regulations of Schedule I.

XVI.—All timber disposed of by the Forest Department shall be stamped with such mark or marks as the Conservator may from time to time direct, and the purchaser and grantees will, on application, receive a pass in the form exhibited in Schedule II.

XVII.—All drift and unclaimed *Sál, Sissoo, Toon, Ebony, Dhao, Assana, Khair,* and *Tikooe* or *Huldoo,* shall be considered the property of Government, unless proof of ownership be given to the satisfaction of the Conservator.

All claims for such drift or timber, the ownership of which is questioned, shall be decided by the Conservator, or such officers as he may authorize to do so; provided, however, that the Forest Officers shall be at liberty to decline to arbitrate regarding such timber in cases where they may see fit to do so, and refer claimants to the civil court.

XVIII.—It is the duty of Forest Officers, and of all Police Officers, to see that these rules are not violated, and should they in any case be infringed, to report the same without delay to the nearest Forest Officer or Magistrate, and when necessary to take into custody, without a warrant, all persons found cutting, marking or removing, without permission, timber from the Government forests.

In cases of an infringement of these rules, other than the cutting, marking or removing timber, it shall be the duty of the subordinate Forest Officers, and of all Police Officers to report the circumstance to the Conservator, or the nearest Magistrate, for orders, or to apply for a summons.

All persons apprehended for the infringement or violation of these rules must be forthwith brought up before a Magistrate—15 miles a day being allowed to reach the court of the nearest Magistrate.

XIX.—(Amended by notification of the Government of India, in the Department of Revenue, Agriculture and Commerce, No. 332, dated 20th September 1871.) In cases in which the penalty of confiscation is not provided by Act VII of 1865, any person who shall infringe any provision of these rules, and any subordinate of the Forest Department who shall wilfully neglect to do his duty under any of these rules, shall be liable to a fine not exceeding Rs. 500, or in default of payment of the same, to imprisonment for a term to be regulated by the provisions of the 67th Section of the Indian Penal Code.

XX.—The Conservator of Forests shall have the powers of a Subordinate Magistrate of the 1st class, but shall exercise these powers only under "the Government Forests Act, 1865," and subject to such limitations as may from time to time be imposed by the Chief Commissioner.

XXI.—The Chief Commissioner may vest any Deputy or Assistant Conservator of Forests with the powers of a Subordinate Magistrate, to be exercised only under the said Act, and subject to such limitations as he may deem proper.

XXII.—A monthly register of all cases tried and determined by each Forest Officer, in any division or district of a Deputy Commissioner, is to be submitted to the appellate court.

The register is to be kept in the form given in Schedule III.

XXIII.—The list of reserved woods comprises :—

Sâl ...	...	...	...	*Shorea robusta.*
Sissoo	...	...	...	*Dalbergia Sissoo.*
Toon	...	...	...	*Cedrela Toona.*
Ebony	...	...	...	*Diospyros Melanoxylon.*
Dhao ...	...	...	...	*Conocarpus latifolia.*
Assana	...	...	...	*Terminalia tomentosa.*
Khair	...	...	...	*Acacia Catechu.*
Tikooe or Huldoo ...	...	...	...	*Nauclea cordifolia.*

XXIV.—Schedule I, containing rules under which Government timber in the Government forests may be disposed of, is to be read and taken as part of these rules.

SCHEDULE I.

1. *By public auction sales at stations or in the forests.*—At all public sales some portion of the payment, not less than 20 per cent. of the amount, is to be made on the day of sale, either in cash, public securities, or promissory notes, otherwise the sale will be cancelled. The balance must be paid within a term not exceeding three months after the date of sale; should the balance not be paid within the specified term, the timber will become the property of Government, and the deposit money be forfeited. No timber is to be delivered before payment in full shall have been received; but the timber shall be at the risk of the purchaser from the moment it is knocked down.

2. *By selection.*—Such selected timber to be paid for at about double the average rate, or at such rate as after a first selection the Conservator may think fair both to the Government and to the purchaser. The conditions as to payment to be similar to those contained in Clause No. 1.

3. *By private sale, or on indent approved of by the Conservator of Forests, or the Chief Forest Officer at the station or in the division in which the sale is held.*—No private sale of Government timber shall be considered as concluded until a payment has been made of 50 per cent. of the purchase money; and no timber shall be removed until the payment of the whole amount has been made; the period for payment to be fixed by the Conservator.

4. *By sale of the old timber standing or lying in a certain forest district, or of living trees of the unreserved woods, to persons undertaking to cut and remove them.*—This timber remains the property of Government until the full amount of purchase money is paid or adjusted. Timber not cut and removed within the period fixed in that behalf by the Government officer selling the same, whether paid for or not, remains the property of Government. Sales of timber under this clause are not to be, ordinarily, concluded for a longer period than two years.

5. *By sale of certain number of trees on indent approved of by the Conservator at specified rates.*—The timber not to be removed from the forest until the amount due for the timber indented for be paid to the Forest Officer in charge of the division. The conditions as to payment to be made by the Conservator or his Assistants.

SCHEDULE II.

Form of Pass of Timber sold by the Forest Department.

Number.	Date.	Owner or Consignee.	Place of Destination.	Description of Timber.	Timber Mark.	Number of logs or pieces.	Total.	Remarks.

SCHEDULE III.

Register of Forest Cases tried and decided by Forest Officers.

Number of Register.	Name, residence, and occupation of accused.	Date and locality when and where the offence is stated to have been committed.	Nature and description of offence.	Nature of Evidence.	Date of Sentence.	Sentence.	Commutation of punishment.	Remarks.

V.—CENTRAL PROVINCES.

NOTIFICATION—By the Government of India, in the Public Works Department, No. 16F., dated 22nd August 1865.

THE following rules for the better management and preservation of the Government forests in the Central Provinces, drawn up under Act VII of 1865, have been confirmed by the Viceroy and Governor General in Council, and are, in accordance with Section 6 of the Act, published in the *Gazette of India :—*

FOREST RULES.—CENTRAL PROVINCES.

I.—The waste lands in the Central Provinces, which are not private property, are all, as respects technical considerations, and as respects timber and other natural products, to be regarded as "Government forests" and to be administered in the Forest Department.

Definition of forests.

II.—These forests are divisible into two classes—(1) "reserved forests;" (2) "unreserved forests." The more valuable tracts will be taken up and declared to be "reserved forests." The remainder, whether reserved from sale under Rule 19 of the Waste Land Sale Rules or liable to sale under those rules, are "unreserved forests."

How classified.

III.—The administration of the forests will be vested in the following officers in the manner hereinafter described :—

How administered

1st.—The Conservator of Forests.
2nd.—His Assistants,
3rd.—The Deputy Commissioners of Districts.
4th.—The Assistant Commissioners.
5th.—The Subordinate Forest Officers, viz., Darogas, Jemadars, Duffadars, Forest Watchers, and Peons.

It will also be the duty of all Police Officers to watch over the observance of these rules, and to afford every assistance to the Forest Officers in the exercise of their duties.

IV.—"Reserved forests," and any others which may be specially assigned to the care of the Conservator of Forests and his Assistants, are managed exclusively by them.

by Forest Officers,

V.—"Unreserved Forests" are under the immediate control and management of District Officers, aided by subordinate Forest Officers, who will be under the orders of the District Officers, but will, as a rule, be appointed by the Conservator of Forests.

by District Officers,

VI.—The boundaries of "reserved forests" will be demarcated by masonry pillars, or in other permanent manner, in concert with Settlement Officers. Proclamation of their having been created Government reserves will be publicly made in the districts.

Reserved forests to be properly demarcated.

(Added by Notification of the Government of India, in the Public Works Department, No. 19F., dated 28th November 1868.) In the reserved forests no land is to be sold, and no lease of land or forest is to be given, without the sanction in every instance of the Government of India.

VII.—Within the limits of "reserved forests," the cutting of any timber, shrubs, and bamboos; the injuring of trees, shrubs, and bamboos; the appropriation of any forest produce; the making of "dhyas;" the grazing of cattle; burning of charcoal; the lighting of fires, or any interference with the ground, or its products, is absolutely prohibited, except with the special permission of the Conservator of Forests. The use of existing roads in "reserved forests" is permitted; but it shall be lawful for the Conservator of Forests, with the sanction of the Chief

and strictly preserved,

Commissioner, to close any road in " reserved forests," and to declare such road to be closed. No person may enter, come out of, or pass through, " reserved forests," except by existing roads; and any person found in a "reserved forest" straying off existing roads, shall be arrested, and shall be liable, on conviction before a Magistrate, to a fine not exceeding Rs. 50, or in default, to rigorous imprisonment for a term not exceeding 14 days.

VIII.—The Conservator of Forests will issue orders and make arrangements for the cutting and sale of timber, and for the disposal of forest produce in " reserved forests." But no *timber trees* shall be cut in such forests, except by the Forest Department itself, which have not been previously marked for cutting by an officer of the Forest Department itself, or which are not included in the tract set apart for felling.

and utilised only under special restrictions.

IX.—" Unreserved forests" will be subject to inspection and to periodical report by the officers of the Forest Department. Those officers may also undertake any operation connected with planting, cutting, thinning, or selling timber in these " unreserved forests," which, in the opinion of the Conservator of Forests, may require their own special attention.

Unreserved forests subject to departmental inspection,

X.—(Amended by Notifications of the Government of India, in the Public Works Department, Nos. 33F. of 1st November, and 35F. of 5th November 1866.) No Teak, *Tectona grandis*; Sâl, *Vatica robusta*; Sâj, *Terminalia glabra*; Sâj, *Terminalia tomentosa*; Beejasâl, *Pterocarpus Marsupium*; Sheshum, *Dalbergia latifolia*; and the Kuttung Bamboo; generally or locally, may be cut in the unreserved forests without the special permission of the Deputy Commissioner, or of the officers of the Forest Department, given in communication with the Deputy Commissioner. But other products in the unreserved forests may be disposed of by the district authorities in the usual manner, provided that nothing be done which might interfere with arrangements of a general or special character made by the officers of the Forest Department. In tracts where posts for the collection of forest duties are not established, and where the right to levy such duties is not leased out, as hereinafter provided in Rule XVII, the rural communities and other residents of a district in the neighbourhood of unreserved forests, will be permitted to appropriate unreserved timber and common jungle products for their own use and consumption, free.

and to be utilised under certain restrictions.

XI.—The district officers, however, may, at discretion, grant permission to cultivators requiring timber of the above-mentioned seven prohibited sorts, *viz.*, Teak, Sâl, Sâj, Beejasâl, Sheshum, Kowah, and Unjun, *bonâ fide* for their own use, and not for purposes of trade; or to local artizans, such as carpenters and others requiring wood for their trades, to cut and remove up to 20 trees yearly. Railway and other contractors, whose timber requirements may be on a large scale, must apply to the officers of the Forest Department with whom will rest the responsibility, subject to the Chief Commissioner's control, of all arrangements with such contractors or other persons who may require large quantities of timber or forest produce.

which may be released in certain cases,

XII.—On timber cut under Rule XI, duty will be charged at the rates laid down in Rule XVI. Cultivators will pay for timber granted them *bonâ fide* for their own use at one-fourth of these rates. This privilege is applicable to cultivators only, and not to local artizans or others. In all cases, the amount of duty must be deposited at the time of making the application. In special cases, grants of timber of the prohibited kinds, free of duty, may be made by the District Officer, with the concurrence of the Conservator of Forests.

on certain conditions,

XIII.—District or Forest Officers granting permission to cut timber under Rule XI must, in every case, depute a person to mark the trees, and such trees shall in no case be of less girth than four and a half feet, at six feet from the ground, and they shall invariably be cut to within one and a half feet of the ground. Sheshum may, however, be cut at three feet girth.

and provisions.

XIV.—No dhya is to be cut in any place where any of the seven interdicted kinds are growing, without the permission, in every instance, of the Deputy or Assistant Commissioner, or Tehsildar of the District. These officers may also give permission for the use of the Sâj leaves, by silkworm rearers, in " unreserved forests."

Concerning dhya cultivation and silkworm rearing.

XV.—Besides exercising the management, as above, of " unreserved forests," District Officers will use every endeavour to give full effect to that clause in the administration papers of proprietary villages in which the proprietors have agreed to adopt the established principles of forest conservancy, as far as may be feasible, in regard to the timber trees on their own lands which are their exclusive property.

Private forests conservancy.

XVI.—The following will be the rates of duty charged on timber in all forests:—

1.—On all logs of Teak, Sheshum, Sâl, Sâj, Kowah, Unjun, and Beejasâl, measuring upwards of eighteen feet, by four and half feet girth at the thick end, Rs. 4 per log.

Rates of duty on forest timber.

2.—On all smaller logs of such timber, Rs. 3 per log.

3.—On all railway sleepers of whatever timber, eight annas per sleeper.

4.—It will be at the option of the Conservator of Forests to charge duty on timber cut under the orders of Forest Officers, at the rate of three annas per cubic foot of squared timber. When such a mode of charge is adopted, the measurement will be made by Forest Officers, and their measurement will be final. The timber will be measured either in the rough or the sawn state.

N. B.—On timber sold by the Forest Department, no duty will be levied.

XVII.—(Amended by notification of the Government of India, in the Public Works Department, No. 16F., dated the 20th October 1867.) In unreserved forests, the realization of Government dues on miscellaneous forest produce and unreserved timber will be regulated as follows :—

In cases where Deputy Commissioners manage these forests by direct agency, duty will be leviable at such rates as the Deputy Commissioner may notify at the commencement of each official year.

In cases where the right to levy such duties is leased to farmers, the lessees will make their own arrangements with their customers, provided that the duties payable for the ensuing year on grass, firewood, rafters (*Mulgahs, Kakas,* and *Kurries*), and the grazing of cattle in transit through wastes, shall be notified publicly for each district by the Deputy Commissioner at the same time when the leases of the district wastes are put up to auction. In no case shall forest duty on the three articles above-mentioned, nor for grazing cattle, be leviable at higher rates than the Deputy Commissioner of the district may have notified to be leviable during the year.

When duties are levied on forest produce borne on rivers, the rates of duty leviable shall be fixed from time to time, by the Conservator of Forests, with the sanction of the Chief Commissioner.

Nothing in these rules shall prevent the disposal of the unreserved sorts of timber or forest produce by sale, or otherwise, as may seem expedient from time to time.

XVIII.—The Conservator of Forests, in granting permission to contractors and others to cut timber in the reserved or unreserved forests, will bind them by such conditions, regarding time and route of removal, method of cutting up the timber into sleepers or scantlings, protection from fire, &c., as he may deem necessary for the prevention of waste, and economizing the supply of timber.

Timber contracts to contain special conditions

XIX. No trees or bamboos are to be cut in any forest, unless by the special direction of District or Forest Officers, within ten yards of the bank of any hill-stream, or within twenty yards of any spring:

XX. The following are to be the bye-laws for the regulation of the timber traffic at timber revenue stations existing, or hereafter to be established, on the Nerbudda or other rivers :—

1.—All persons purposing to bring timber down such river must register, in the office of the Assistant Conservator of Forests, a specimen or a *facsimile* of the brand wherewith their timber shall be marked.

2.—Government timber will be marked with such mark or marks as the Chief Commissioner may, from time to time, direct, to distinguish it from all others.

3.—All timber without exception, that comes down such river, will be caught in the river by persons employed by the Forest Department, and lodged by them within the Government timber-yard.

4.—The timber will be given over to the owners, when not disputed, on payment of the authorized amounts for catching and lodging the timber in the timber-yard, in addition to the regular timber duty. The Chief Commissioner will, from time to time, fix the amounts to be paid, and will prescribe a form of receipt to be granted to the Forest Officers in charge of such stations. Should the amount of fees and duty not be paid, and the timber removed within one month, it will be sold, and the proceeds credited to Government. Should there be any dispute, the timber will be kept in the yard until the ownership is settled in the civil court, provided that a suit be filed by one of the claimants within one month.

5.—All timber found in the river without a registered brand or mark will be confiscated to Government.

XXI.—All persons receiving permission to cut large quantities of timber in Government forests, will be required to register, in the office of the Assistant Conservator of Forests of the division, a specimen or *facsimile* of the brand or mark wherewith their timber shall be marked under the superintendence of the forest officials. No private parties will be allowed to use the Government mark, or any other mark already registered. A fee of Rs. 10 will be charged for registration of property-marks for timber. Any person found using any mark other than his own registered mark, or effacing any property-mark lying within forest limits, or floating on a river before the said timber has paid duty, will be liable, on conviction before a Magistrate, to a fine not exceeding Rs. 500, or, in default of payment, to imprisonment of either description as defined in the Indian Penal Code, for a term not exceeding six months.

XXII.—(Amended by notification of the Government of India, in the Department of Revenue, Agriculture, and Commerce, No. 40, dated the 9th January 1874.) Any person who shall within the limits of any Government forest, cut or convert to his own use, lop, mark or injure any trees, shrubs or bamboos, or who shall make any dbyas, burn charcoal, light fires, graze cattle, or do any act endangering the safety of the forest, except in pursuance of a right conferred by these rules, or in pursuance of the written permission of the officers authorized to grant such permission, shall, in cases where the penalty of confiscation is not provided by the Act, be liable, on conviction before a Magistrate, to a fine not exceeding Rs. 500, or, in default of payment of such fine, to imprisonment for such term as is mentioned in Section 67 of the Indian Penal Code.

XXIII.—No Forest Officer shall engage in any other employment or office whatever other than his duties under these rules, unless expressly authorized to do so in writing by the Conservator of Forests.

XXIV.—Any subordinate Forest Officer, who shall be guilty of any violation of duty, or neglect of any rule or regulation, or lawful order, made by a competent authority, for his guidance in matters connected with guarding the boundaries of the forests; the marking, girdling, or felling of trees; the marking and passing of timber; the reporting and preventing of offences against the forest rules; or who shall engage without authority in any employment other than his forest duty; or who shall withdraw from the duties of his office without permission, or without having given previous notice for the period of two months, shall be liable, on conviction before a Magistrate, to a penalty not exceeding Rs. 250, or, in default of payment, to simple imprisonment for a term not exceeding three months.

XXV. All contracts, licenses, or permissions granted to parties to exercise any privilege, or to do any act in forests under the preceding rules, must contain a condition that infringement of any general rule, or special provision, will entail forfeiture of such contract, license, or permission, and also of any wood or forest product cut or gathered, and still within forest limits.

General conditions applicable to all contracts or permissions.

VI.—COORG.

NOTIFICATION—By the Government of India, in the Department of Revenue, Agriculture, and Commerce, No. 129, dated Simla, the 11th August 1871.

The following rules, drawn up under Act VII of 1865, have been confirmed by the Viceroy and Governor General in Council, and are, in accordance with Section 6 of that Act, published in the *Gazette of India :—*

RULES FOR THE BETTER MANAGEMENT AND PRESERVATION OF THE GOVERNMENT FORESTS IN COORG.

I.—The following rules are published for the administration of such Government forests in the Province of Coorg as have been defined in notification No. 127 of the 10th instant.

Forest rules: their object.

II.—The administration of these forests will be vested in the following officers:—

Officers appointed for their administration.

(a.)—The Conservator of Forests, his Assistants, and the subordinate Forest Officers.
(b.)—The Superintendent of Coorg and the subordinate Revenue Officers. It will also be the duty of all Police Officers to watch over the observance of these rules, and to afford every assistance to the Forest Officers in the exercise of their duties.

III.—The boundaries of Government forests will, wherever they do not run along a road or stream, or other well-defined line, be demarcated by cleared boundary paths and permanent boundary marks. Wherever practicable, the boundary lines of Government forests and the boundary marks should be entered on maps which should be prepared in triplicate; one copy to be sent to the Conservator of Forests, one copy to remain with the Forest Officer in charge of the division, the other to be deposited in the office of the Superintendent of Coorg. In special cases, the Chief Commissioner of Coorg may authorize the demarcation of State forests by natural boundaries without cleared paths and without artificial boundary marks. In like manner, the Chief Commissioner may dispense with the preparation of the map showing the boundary lines as here prescribed.

Government forests how demarcated.

Proclamation of the demarcation of Government forests and their boundaries will be publicly made in the talook where they are situated, and notified in the *Mysore Gazette*.

IV.—In Government forests no land should be alienated without the orders of the Government of India.

In Government forests no land to be alienated.

V.—The Government forests will be under the exclusive control of the officers of the Forest Department. Unauthorized felling, cutting, lopping, marking, or injuring of trees, shrubs or bamboos, or the collection of leaves, grass, gums, resins, and other forest produce, the clearing of land for kumri cultivation, the setting fire to grass or jungle, the grazing of cattle, or any act that is likely to damage the forests, is prohibited, and will be punished by fine not exceeding five hundred rupees, and in default of payment of such fine by imprisonment for such term as is provided in the 67th Section of the Indian Penal Code.

Government forests to be strictly preserved.

VI.—Existing roads or pathways through the Government forests may be used as far as is compatible with the conservancy of the forests; but the Conservator of Forests, with the concurrence of the Superintendent of the Province, may close any existing roads or pathways through any Government forests, whenever he may deem it requisite to do so. Public notice of the closing of such a road shall be given in the talook or talooks where the forest is situated.

Trespass in Government forests.

Ingress to the Government forests without permission, except by authorised roads and footpaths, is prohibited. Any one found off the authorized roads and footpaths in the forests without authority, and owners of cattle straying in the forests, will be liable to a fine not exceeding two hundred and fifty rupees, and, in default of payment of such fine, to imprisonment for such terms as is prescribed in the 67th Section of the Indian Penal Code.

Cattle found straying in the forests may be pounded, and may be redeemed on payment of a sum of money according to a scale of rates to be laid down from time to time by the Chief Commissioner of Coorg, and, in default of payment of such sum of money within a reasonable time, the cattle shall be sold on account of Government. It shall be lawful for the officer selling such cattle to award a portion of the proceeds of such sale, not exceeding one-half, to any person on whose information such cattle was seized. Such fines to be credited to the Forest Department.

VII.—There will be a subordinate Forest Officer in charge of every Government forest or part of a Government forest. He must reside within or in the immediate vicinity of the forest. He must be acquainted with every part of it, and with whatsoever happens therein. He will be responsible for the maintenance of the boundary lines and boundary marks.

Officers in charge of Government forests.

In cases of unauthorized felling and other breaches of the forest rules, he must immediately report the occurrence.

It will further be his duty by all means in his power to prevent the continuance or repetition of the acts constituting the breach. He will seize all wood or other forest produce unlawfully cut or removed, which he may find within the limits of the forest.

He will use every lawful means for the defence of the property entrusted to his charge.

He will be held responsible that no trees, except those marked by the Conservator or his Assistants, are felled.

VIII.—No Forest Officer shall engage in any employment or office whatsoever other than his duties under these rules, unless expressly permitted to do so in writing by the Chief Commissioner of Coorg.

Forest Officers not to engage in other employ.

IX.—All drift and unclaimed timber and bamboos within the Province of Coorg will be considered the property of Government, unless proof of ownership be given as hereinafter provided. Drift timber and bamboos shall be collected at such stations as the Conservator of Forests may direct, and notices shall from time to time be published, stating the number and description of pieces of drift timber and bamboos collected at such stations.

Drift and unclaimed timber, the property of Government.

X.—Not less than two months' notice will be given for the reception of claims to the ownership of drift and unclaimed timber or bamboos, after which no claims will be allowed, and the timber and bamboos will be sold on account of Government.

Notices inviting claimants.

XI.—All such claims will be settled by the Conservator, or by such officer as he may authorize: provided, however, that he shall be at liberty to decline arbitrating regarding such timber or bamboos, and, in case he may see fit to do so, refer claimants to the civil courts.

Claims by whom decided.

XII.—Timber or bamboos awarded to claimants must be redeemed by the payment of salvage and other expenses which may have been incurred on account of such timber.

Claimed timber how redeemed.

XIII.—It is the duty of the officers and subordinates of the Forest Department, and of all Revenue and Police Officers, to see that these rules are not violated, and, should they in any case be infringed, to report the same without delay to the nearest Magistrate or Forest Officer in charge of the range; and it shall be lawful for any Forest Police Officer to take into custody, without a warrant, any person who hinders or obstructs him in the discharge of his duties under these rules, and the person apprehended shall be brought before a Magistrate with the least possible delay.

Duties of Forest, Revenue, and Police Officers with regard to these rules.

XIV.—Any person who infringes any provision of these Rules for which no special penalties have been provided, or any subordinate Forest Officer who wilfully neglects his duty, will be liable to a fine not exceeding five hundred rupees, and, in default of payment, to imprisonment for such term as is prescribed by the 67th Section of the Indian Penal Code. In cases where the infringement involves fraud or theft, or any other offence provided in the Penal Code, the offender will be liable to be proceeded against under the provisions of the Penal Code.

Penalty for breach of forest rules.

XV.—Any axes, knives, carts, boats, or other tools, vehicles, or implements, as also all cattle and domestic animals used in an act which constitutes an offence against these rules, and all timber which has been marked or obtained in a manner contrary to these rules, whether entire or cut up, or sawn up, may be seized, by any officer of the Forest

Tools, timber, and other articles may be confiscated.

Department, or Police Officer; and such timber, tools, vehicles, implements, cattle and domestic animals may be confiscated or released on payment of a fine by the orders of the Magistrate of the district, or of such Magistrate, Subordinate Magistrate, or Forest Officer vested with any of the powers of a Magistrate, as may be specially empowered by the Chief Commissioner to exercise jurisdiction under these rules.

XVI.—Offences against these rules may be tried and determined by the Magistrate of the district, or by such Magistrate, Subordinate Magistrate or Forest Officer vested with any of the powers of a Magistrate, as may be specially empowered by the Chief Commissioner to exercise jurisdiction under these rules, provided that Magistrates, Subordinate Magistrates, or Forest Officers, vested with the powers of a Magistrate, shall not exceed their respective powers as defined in Section 22 of the Code of Criminal Procedure.

Cases of violation of the rules by whom tried.

XVII.—The Chief Commissioner may invest the Conservator, Deputy Conservator or any Assistant Conservator of Forests, who may be qualified, with the powers of a Magistrate or of a Subordinate Magistrate under these rules.

Forest Officers may be vested with magisterial powers.

XVIII.—The Chief Commissioner of Coorg shall be at liberty to frame rules and revise such rules from time to time as shall be necessary for the sale by auction or otherwise of sandal wood, timber, or any other forest product, produced in the forests. Such rules shall be binding on all purchasers, and any breach in their observance shall render the offender liable, on conviction, to the penalties detailed in Sections 14 and 15 of these rules.

VII.—BENGAL.

NOTIFICATION—By the Government of India, in the Public Works Department, No. 13P., dated 18th February 1871.

THE following rules, drawn up under Act VII of 1865, have been confirmed by the Viceroy and Governor General in Council, and are, in accordance with Section 6 of the Act, published in the *Gazette of India:—*

RULES FOR THE BETTER MANAGEMENT AND PRESERVATION OF THE GOVERNMENT FORESTS IN THE LOWER PROVINCES OF BENGAL.

PART I.

Preliminary.

These rules shall be in force in those tracts of land which may be declared to be Government forests under Act VII of 1865, and they shall regulate the transit of timber and other matters in those districts to which these rules may be extended by an order of the Government of Bengal published in the Gazette.

2. On any tract being declared by notification under Section 2, Act VII of 1865, to be subject to the provisions of that Act, it shall be subject to the authority of the Conservator of Forests under the following rules. Provided that the Government may appoint any person, other than the Conservator of Bengal Forests, to be Conservator of any specified tract.

3. Divisions and sections of Government forests will be placed under the management of Deputy and Assistant Conservators, as may be determined by the Government of Bengal.

4. All officers charged with the protection and management of any Government forests, or with the control and management of any matters to which these rules relate, will be designated Forest Officers.

5. The Government forests, as notified under Section 2, Act VII of 1865, shall be marked on a copy of the revenue survey map of the district, which shall be deposited in the office of the Commissioner of the division, and attested by his signature.

6. There will be two classes of forests—"reserved" and "open"; and whenever any notification is issued under Section 2 of Act VII of 1865, rendering any land subject to the provisions of that Act, such notification shall state whether such land is to be a reserved or an open forest.

PART II.

Of Reserved Forests.

7. The reserved forests shall consist of tracts or forest of waste lands, the soil of which is the property, and is in the possession, of Government, which may be set apart as such by order of Government, and defined by notification in the Official Gazette. The Conservator of Forests and his subordinates shall have entire administration, custody and control over these forests and their products.

8. Within reserved forests no tree of any kind is to be felled, nor shall any forest produce be removed without general or special authority obtained from the Conservator of Forests.

9. The boundaries of these forests must be clearly marked off by substantial marks where no natural boundaries exist. Should villages be included within these tracts, the boundaries around the village lands must be demarcated in the same manner as the forest, subject to the proviso in Section 2, Act VII of 1865.

10. Whenever the Government may resolve to set apart any tract as a reserved forest, the privileges which villagers are to be allowed to enjoy, and any customary rights which may belong to them in the matter of providing themselves with firewood and wood for domestic and farming purposes and otherwise, shall be distinctly defined and approved by Government. It shall not be competent to the Conservator in any way to restrict or modify privileges so allowed, except with the approval of Government.

11. As soon as the notification required by Section 2 of the Government Forest Act, stating that the forest is to be a reserved forest, is published in the Gazette, a notification shall be published and issued to the villages around, mentioning the boundaries of the tract reserved, warning the villagers against trespass or mischief, and defining the restrictions imposed on them, and the privileges allowed to them, within the reserved forest under this section. This notification will be issued by the District Officer, and a copy will be sent to the Forest Officer for information. The Conservator will, nevertheless, be careful to give prompt attention and consideration to any complaints which may subsequently be preferred to the effect that the restrictions so imposed infringe rights of, or inflict hardship upon, individuals or communities.

12. The Conservator of Forests may prohibit all persons from passing through a reserved forest, except by the authorized roads or paths, of which a list shall be published, drawn up by the Conservator in conjunction with the chief civil authority.

Part III.

Of Open Forests.

13. The open forests shall consist of such tracts of forest or waste land notified under Section 2, Act VII of 1865, as may not by special notification be included in the reserved forests. In these tracts the authority of the Forest Department shall extend only to the protection of such reserved trees as may from time to time be notified with the approval of the Local Government.

14. Within the limits of open forests, and subject to the proviso in Section 2, Act VII of 1865, no person, without the permission of the Forest Officer, may mark, cut, girdle, or fell, or in any way injure any tree of the kinds which may be reserved as aforesaid.

15. Until the Government forests, to which these rules shall be applicable in any district, are settled and defined under Section 2 of the Act, no lands in the district which are covered with trees, brush-wood, or jungle, and after the forests shall have been defined, no such land in the open forests, shall be sold, nor shall grants or leases in such lands be given, except under the orders of the Commissioner, who should, before passing final orders, communicate with the Forest Department; and, in case of difference of opinion between him and that Department, refer the question to the Government.

16. In addition to the reservation of the trees of the kinds which may be notified as reserved under Rule 13, it shall be competent to the Conservator of Forests and his Assistants to prohibit the felling, cutting, or otherwise using trees of other kinds which they may have marked or girdled within the open forests.

17. No other restrictions shall be imposed by the Forest Department in regard to the open forests, except after the publication of notifications with the approval of the Government.

18. Should the Conservator think it desirable to impose any further restrictions than those above specified in any open forest, he shall, in the first place, communicate his proposals to the Commissioner of the division, who shall give his opinion on the subject. Before approving of any such proposals, it will be the duty of the Commissioner of the division to satisfy himself that the proposed restrictions will infringe no rights or privileges which should be preserved, and to define precisely any privileges which should be kept intact. On receipt of the proposals as agreed upon between the Conservator and Commissioner, the Government will sanction the issue of notifications imposing such restrictions as it may see fit.

Part IV.

Of the use of streams, of marking-hammers, of timber in transit, and of drift timber.

19. A list of rivers, streams and waters in each district to which these rules may have been extended, which must be kept open for the passage of timber, will be made by the Forest Department in conjunction with the civil officers of the district, and be published for general information; and it will be the duty of the civil officers to see that these streams, rivers, or waters are kept open.

20. No timber, the produce of any Government forest, found adrift or stranded in any river, stream, or water of any district to which these rules may be extended, shall be marked until disposed of by the Forest Department; neither shall any mark on it be defaced. It shall not be converted or cut into pieces, neither shall it be removed or disposed of in any way without the orders of the Forest Officer, except on an order or decree by a competent court.

21. All timber disposed of by the Forest Department must be stamped by the Forest Department with such marks as the Conservator may direct.

22. All brands, or marking-hammers used in the marking of timber, must be registered in the principal forest office of the division in which they are to be used. A fee of Rs. 10 must be paid for every mark registered, and a certificate will be granted on payment of fee.

23. The use by any private person of any Government timber mark, or of any other timber brand or mark, save the one registered in his own name, is prohibited.

24. All foreign timber brought into British territory shall be reported and stopped at such stations as the Government of Bengal may from time to time direct, and such timber shall not be allowed to pass until it has been examined. The persons in charge will receive a pass in the form shown in Schedule A, and the timber will be liable to detention, if found afterwards without a pass in any district to which these rules are extended.

25. All timber found adrift or stranded in the rivers, streams and waters of any district to which these rules may be extended, and which is the produce of any Government forest, will be considered the property of Government, unless proof of ownership be given.

26. Persons who may have saved such drift timber are required to deliver the same to such persons as are authorized by the Conservator of Forests to receive it, on receipt of such salvage rates as may be approved by the Government of Bengal.

27. The Forest Officers shall take possession of any drift timber found in any river or stream, or lying in the bed of any river or stream within any district to which these rules may be extended, as to the ownership of which there may be any doubt; and the same shall be treated as drift timber under Rules Nos. 28, 29, 30, 31 and 32 below.

28. All drift timber collected by Forest Officers shall be stored at such depôts as the Conservator of Forests may from time to time order to be established.

29. From time to time, as the collection of drift timber may render it advisable, public notice, inviting claimants to the ownership of such drift timber to come forward, will be issued at the chief stations of the district, and in the nearest town to the place or depôt at which such drift timber may be lying. These notices shall state the number and description of logs collected.

30. Six months' notice will be given for the reception by the Forest Officer, by whom the notice was issued, of claims to drift timber, after which, if not claimed, the timber will be sold on behalf of Government.

31. Claims to drift timber must be sent to the Forest Officer by whom the notice was issued, with particulars of marks by which it may be recognised. The said Forest Officer will inquire into the claim. If the claim be a single one, he may, on being satisfied of its validity, release the timber after expiration of the term mentioned in the notice. If the claim be a disputed one, he should endeavour to obtain the formal consent of the claimants to his arbitrating on their claims. Where the dispute cannot be so decided, and there exists any reasonable doubt as to the rightful owner, the Forest Officer will retain possession of the timber until the question of ownership be decided by the civil court having jurisdiction in the case: Provided that if no claim be preferred in such court within two months from the date on which the Forest Officer shall declare his inability to settle the dispute, the timber may be disposed of by the Forest Officer on account of Government.

32. Timber awarded to claimants must be redeemed by payment of the salvage rates approved by the Government of Bengal, and other expenses which may have been incurred on account of such timber.

PART V.
Of the prevention of offences, and of punishments.

33. It shall be the duty of all Forest Officers to see that these rules are not violated; and should they in any case be infringed, to arrest the offenders and convey them without delay to the nearest Magistrate or Police Officer in the division or sub-division in which the offence took place.

34. Any person infringing any of these rules shall be liable to be punished by fine not exceeding Rs. 500, or, in default of payment of such fine, by imprisonment for such term as is mentioned in the 6th Section of the Indian Penal Code. But no fine shall be imposed in addition to the confiscation of any timber or other forest produce, or of implements, which may have been incurred under Act VII of 1865.

35. Any marking-hammers or other tools or implements, boats, carts and cattle used in an act which constitutes an offence against these rules, and all timber that has been marked or obtained in a manner contrary to these rules, or that has not been reported and passed in accordance with these rules, whether entire, or cut up, or sawn up, may be seized by an officer of the Forest Department, or Police Officer, and such tools or implements, boats, carts, cattle and timber may be confiscated by the orders of the Magistrate of the district.

SCHEDULE A.
Form of Pass for foreign timber imported into British Territory.

Number.	Date.	Owner or consignee.	Place of destination.	Description of timber.	Timber marks.	Number of logs or pieces.	Total.

VIII.—BRITISH BURMA.

A.—FOREST RULES OF 1865.

NOTIFICATION—By the Government of India, in the Public Works Department, No. 15F., dated 2nd August 1865.

THE following rules for the better management and preservation of the Government forests in British Burma, drawn up under Act VII of 1865, have been confirmed by the Viceroy and Governor General in Council, and are, in accordance with Section 6 of the Act, published in the *Gazette of India* :—

RULES FOR THE ADMINISTRATION OF FORESTS IN BRITISH BURMA.

CHAPTER I.

Introduction.

The following rules are published for the administration of the Government forests in British Burma, and for the management of foreign and drift timber :—

Forest rules—their object.

Limits of the forests. 2. The Government forests of British Burma are bounded and limited as follows :—

FIRST.—*Forests between the Irrawaddee River and the range of hills east of the Sittang River.*

These forests comprise the hills and valleys of the Pegu Yomah range, and are bounded as follows :—

On the west by a line passing from the station of Meaday through the points of junction of the following streams : —

Boolay and Pudday,
North and Middle Nawing,

thence to the head-waters of the Meimakah river east of Prome, and along that river to the mouth of the Meoungdagah Khyoung in the Rangoon district.

On the south by a line drawn from the mouth of the Meoungdagah Khyoung across to Pegu town.

On the east by a line drawn from Pegu town to Bonee village, thence to Kanyinokdoh village between the Binedah and Yainway streams, thence to the junction of the Woon and Koon streams, and thence across the Sittang river to Kyonkee town, and to the hills between the Sittang and Salween rivers, and along the line of water-shed between the Sittang and Salween rivers, to the northern frontier line.

SECOND.—*The Forests in the Tenasserim Division of British Burma.*

1. The Beeling forests above the village Phawota on the Beeling river.
2. The Yoonzaleen forests above the village of Kadinetitate on the Yoonzaleen river.
3. The Domadamee forests above the junction of the Domdamee stream with the Salween river.
4. The Thoungyeen forests between the hills bounding the Thoungyeen valley on the south-west, and the Siamese boundary line.
5. The Attaran forest above the junction of the Zimmay and Winyeo streams.

Officers appointed to carry out these rules. 3. The officers appointed for the administration of the forests are—

1. The Conservator of Forests.
2. His Assistants, *vis.*, the Deputy and Assistant Conservators, and the Native Forest Assistants.
3. The Subordinate Officers, *vis.*, Goungs, Goungways, and Peons, both in the forests and at the timber stations.

The administration of the forests throughout the whole province of British Burma shall be vested in the Conservator of Forests.

The administration of one division or sub-division of the forests shall be vested in a Deputy or Assistant Conservator.

4. There will be a subordinate Forest Officer in charge of every Government forest district. He must live within or close to the forest.

Duties of subordinate local Forest Officers. He will be expected to be thoroughly acquainted with every part of the forest, and whatever happens therein. In cases of unauthorized felling or other breach of the forest rules, it will be his duty to report the occurrence immediately. If will further be his duty to prevent, by all means in his power, the continuance or repetition of the acts constituting the breach. He will seize all wood, or other forest produce, unlawfully cut or removed, which he may find in the forests. He will use every lawful means for the defence of the property entrusted to his charge. He will be held responsible that no teak trees, but those selected by the Conservator or his Assistants, are felled.

5. No Forest Officer shall engage in any employment or office whatever other than his duties under these rules, unless expressly permitted to do so in writing by the Conservator of Forests.

Forest Officers not to engage in other employ.

CHAPTER II.

Of teak trees and trees of other kinds.

6. No person is permitted, without orders from the Conservator or his Assistants, to mark, girdle, or fell any teak trees, large or small, to cut or to break off branches from teak trees, or to injure them in any way.

Teak trees—how protected.

7. The felling and dragging of timber must be done in such a manner as not to break or injure any teak trees. Owners of elephants are responsible for any injury done by their animals.

Precaution to be taken in felling and dragging of timber.

8. No Government teak forests may be sold, and no grant of land may be given within the limits of a Government teak forest, without the special sanction, in every instance, of the Government of India, with the exception mentioned below. Nor shall any lease be given for any Government teak forest, which includes the permission to girdle or otherwise to kill or fell green teak trees, without the special sanction, in every instance, of the Government of India. In forests thus leased, grants of waste land may be given without the special sanction of the Government of India. But the grantee will be bound, over and above the price of the land, to pay to the lessee of the forests, an indemnification for every tree above a certain size standing on the land at the time the grant is made. The rate of this indemnification to be the same as the rate paid to Government by the lessee for the timber under his lease.

Sanction of Government of India required for sale and lease of forests.

9. No Toungya is to be formed on any spot on which stand teak trees, large or small, green or seasoned, without the permission of the officer in charge of the forest district. It shall be lawful for the Chief Commissioner to exempt certain forest districts from the operation of this rule, and also to cancel the order of exemption at any time.

Toungyas where prohibited.

10. All teak timber which may be lying in a place selected for a Toungya, must be protected against destruction by fire. The cultivators of the Toungya will be held responsible for all injury done to such teak timber by fire.

Teak in Toungyas to be protected from fire.

11. Whoever sets fire to any teak forest, or causes the conflagration of a teak forest, is liable to punishment under the Indian Penal Code. Travellers will be held responsible for the extinction of their camp fires, and Toungya cultivators must, to the best of their ability, prevent the spread of Toungya fires into teak localities.

Setting fire to teak forests prohibited.

12. All trees (except teak) as also bamboos are free. But it shall be lawful for the Chief Commissioner to authorize the levying of duty on the felling, cutting, or otherwise using of trees of other kinds; as also of prohibiting these being felled, cut, tapped, or in any way made use of, if they are below a certain size. It shall also be lawful for the Conservator of Forests or his Assistants to prohibit the felling, cutting, or otherwise using of any trees which they may have marked or girdled.

Other trees and bamboos are free, with certain exceptions.

CHAPTER III.

Of Reserved Forests.

13. Certain tracts of forest land or waste will be reserved as the exclusive property of the State. Such waste or forest lands will be termed *" Government reserved forests."* Unauthorized felling, cutting, marking, killing, or injury of trees, shrubs or bamboos, or the collection of leaves, wood oil, resin or other forest produce within the limits of the Government reserved forests, or any act which violates the rights of the State as proprietor of these forest lands, will be punished. Toungya cultivation, and interference of any kind with the ground or its produce, without the special permission of the officers in charge of the Government reserved forests, is prohibited. The right of using existing roads through the Government reserved forests, will be free if compatible with the conservancy of the forests. But the Conservator of Forests, with the sanction of the Chief Commissioner, shall be at liberty to close any existing roads through a reserved Government forest, wherever he may deem it requisite to do so.

Reserved tracts defined.

14. The boundaries of each Government reserved forest will be marked off as the Chief Commissioner may direct. Where the boundary is not an established road, the bed of a river or other line easily traced, a path at least six feet wide must be cleared through the jungle along the whole boundary line, and the marks should not be more than 200 yards apart. A proclamation must be issued and published in the nearest villages within a month of the final demarcation of the forest, stating in a general way the boundaries of the forest, and warning against trespass, theft, or mischief. It shall ordinarily be the duty of the Deputy Commissioner of the district to issue this proclamation. A copy of it is to be deposited in the office of the Deputy Conservator of the district.

Demarcation and boundaries of Government reserved tracts.

The officer in charge of a Government reserved forest must, twice a year after the close and before the commencement of the rains, go along the whole of his boundary lines, and state in his report the state of the boundary lines and marks.

It is the duty of the officer in charge of a "Government reserved forest" to report immediately every encroachment of the boundaries of his forest.

15. Should villages be situated so as to be surrounded on all sides by a Government reserved forest, the boundaries around their lands must be demarcated in the same manner as the boundaries of a Government reserved forest; or should the villages be merely temporary settlements, they may be allowed to remain for a period not exceeding three years, under such conditions as the Conservator of Forests, and the Deputy Commissioner of the district may determine under approval of the Chief Commissioner.

Demarcation of villages within reserved tracts.

But no cultivation, by clearing or burning the jungle, is, on any account, to be permitted within the boundaries of a Government reserved forest.

16. The Conservator of Forests, or any Deputy Conservator in charge of a forest division, may reserve any tract in the forests not exceeding 100 acres, provided such tract do not contain houses or cultivation. Forests of a larger size can only be reserved by the Chief Commissioner. The procedure to be observed in the reservation of forest tracts is described in Schedule I. The Chief Commissioner may at any time sanction alterations in the rules of this Schedule.

Reserved tracts procedure.

CHAPTER IV.

Of the use of Streams and Canals, of Marking-Hammers, and Management of Drift Timber.

17. The right of floating timber by natural streams and artificial canals in British Burma is reserved in all cases, subject to the control of the District Officer and other Government authorities. A list of streams in each district which may not be closed, nor blocked up, neither partially nor entirely, neither for fisheries, irrigation, nor for other purposes, is given in Schedule II. The Chief Commissioner may at any time add streams to this list. The throwing in of Toungya refuse into these streams is prohibited.

Streams and canals not to be blocked up.

18. The Chief Commissioner may prohibit the poisoning of streams for the purpose of fishing in any division of the forests.

Poisoning of streams may be prohibited.

19. No teak timber which is subject to the control of the Forest Department, or found adrift in the creeks and rivers of the country, shall be marked, neither shall any mark on it be effaced. It shall not be converted, cut into pieces, nor burnt; neither shall it be concealed, removed, nor disposed of by sale or otherwise without orders from the Conservator of Forests or his Assistants

Teak timber—how protected.

20. The use of marking-hammers, except by an order of an officer of the Forest Department, on the rivers enumerated in Schedule III, and on such other rivers as the Chief Commissioner may direct, is prohibited.

The marking of timber—where prohibited.

21. The disposal of timber from the Government forests of British Burma, either by sale or gratuitously for the common public benefit, will take place according to the regulations of Schedule IV. The Chief Commissioner may, from time to time, amend these regulations.

Government teak timber—how to be disposed of.

22. All timber disposed of by the Forest Department will be stamped with such mark or marks as the Conservator of Forests may, from time to time, direct; and the purchasers or grantees will, on application, receive a pass in the form exhibited in Schedule V.

Marks and passes for Government timber sold.

23. Foreign teak timber, when brought across the frontier on the Irrawaddee and Sittang rivers, must be reported at such stations as the Chief Commissioner may, from time to time, direct. Such timber will receive a pass in the form exhibited in Schedule VI.

Foreign teak timber—how to be dealt with on the Irrawaddee and Sittang rivers.

24. All teak timber, which is brought down the Salween, Beeling, Attaran, or any other rivers to Moulmein, whether from beyond the frontier or from the forests in British territory, must be reported and passed in the manner described in Schedule VII.

Teak timber brought to Moulmein—how to be dealt with.

25. All teak timber, which is brought to Rangoon or Bassein, will be examined at the river stations named in Schedule VIII. No rafts of timber shall leave or pass these stations without an order from the Forest Department.

Teak timber brought to Rangoon and Bassein.

26. All drift and unclaimed teak timber within the Province of British Burma, will be considered the property of Government, unless proof of ownership be given according to Rules 31, 32, and the provisions of Schedule VII.

Drift timber—the property of Government.

27. Scales of salvage for the different parts of the country are given in Schedule IX.

Scales of salvage for drift timber.

They may be altered by the Chief Commissioner. Notifications exhibiting these rates will, from time to time, be published for general information.

28. Permission to parties wishing to collect their own timber which may have gone

Permission to save drift timber—how given.

adrift, will be granted on application at the discretion of the Conservator of Forests, or such Assistants as he may authorize to give this permission. The permit will be made out according to form in Schedule X. These permits must show the marks which should be on the timber, and the holders will only be permitted to take away such logs as bear these marks.

29. Parties who have saved teak timber are bound to deliver the same, on receipt of

Parties who have saved drift timber—how to act.

salvage money, to such persons as are authorized by the Conservator of Forests to receive it.

30. At the stations of Rangoon, Prome, Tounghoo, and Moulmein, and at such other

Monthly notices of drift timber brought in, to be issued.

stations as the Conservator of Forests may direct, notices shall be published on the last day of every month, stating the number and description of pieces of drift timber brought in during the month.

31. Not less than 30 days' notice will be given for the reception of claims to the ownership

Notices inviting claimants to come forward.

of drift timber by the office from which the notice was issued, after which, no claimant appearing, the timber, with such exceptions as are provided for in Schedule VII, will be sold on account of Government. Claims for drift timber must be sent in with full particulars, and according to form given in Schedule II.

32. All such claims will be decided by the Conservator of Forests, or such officers as he

Claims of drift timber—how decided.

may authorize so to do, provided, however, that they shall be at liberty to decline to arbitrate regarding such timber, in case they may see fit so to do, and refer claimants to the civil courts.

33. Timber awarded to claimants must be redeemed by payment of the salvage and other

Claimed timber—how redeemed.

expenses which may have been incurred on account of such timber.

CHAPTER V.

Procedure and Penalties in case of a breach of these Rules.

34. It is the duty of the forest goungs, goungways, and peons of the Department,

Duty of Forest Department subordinates to report breach of these rules.

and of all Police Officers, to see that these rules are not violated; and should they, in any case, be infringed, to report the same without delay to the Forest Officer in charge of the division or sub-division in which the offence took place; and it shall be lawful for any Forest or Police Officer to take into custody, without a warrant, any person who hinders or obstructs him in the discharge of his duties under these rules, provided that the person thus apprehended be brought before a Magistrate within three days from the date of his apprehension.

35. Any person who infringes any provisions of the forest rules, or any subordinate

Penalty for breach of forest rules.

of the Forest Department who wilfully neglects his duty, will be liable to imprisonment without labour for a term not exceeding six months, or to a fine not exceeding Rs. 200, commutable, if not paid, to imprisonment for a term not exceeding six months without labour. In cases where the infringement involves fraud or theft, or any other offence provided in the Penal Code, the offender will be liable to be proceeded against in a criminal court, under the provisions of the Penal Code.

36. Any axes, dahs, marking-hammers, or other tools or implements used in an act which

Tools, timber, and other articles may be confiscated.

constitutes an offence against these rules, and all timber that has been marked or obtained in a manner contrary to these rules, or that has not been reported and passed in accordance with Rules 21 to 24, and Schedules V to VIII, whether entire, or cut up, or sawn up, may be seized by any officer of the Forest Department or Police Officer, and such tools, or implements, and timber may be confiscated by the orders of the Magistrate of the district, or any Forest Officer exercising the powers of a Magistrate, or a Subordinate Magistrate.

37. The Conservator of Forests shall have the powers of a Subordinate Magistrate of

Powers of the Conservator of Forests under these rules.

the 1st class, but shall exercise those powers subject to such limitations as may, from time to time, be imposed by the Chief Commissioner.

38. The Chief Commissioner may vest any Deputy or Assistant Conservator of Forests

Powers of the Deputy and Assistant Conservators.

with the powers of a Subordinate Magistrate of the 1st class, subject to such limitation as he deems proper.

39. All cases of violation of these rules may be tried and determined by the Magistrate

Procedure in cases of violation of these rules. of the district, by the Conservator of Forests, or any Deputy or Assistant Conservator who may have been vested with the powers of a Magistrate, or a Subordinate Magistrate. The mode of procedure in the trial of forest cases will be that provided in Act XXV of 1861.

40. The appeals against the decision of the Conservator of Forests, while exercising the

Appeals where to lie. powers of a Magistrate under these rules, will lie to the Court of the Commissioner of the division where the offence was committed. And the appeal against any decision of a Deputy or Assistant Conservator of Forests, acting as a Magistrate under these rules, will lie to the Court of the Deputy Commissioner of the district.

A monthly register of all cases tried and determined by each Forest Officer in any division or district of a Deputy Commissioner, is to be submitted to the appellate court. The register is to be kept in the form given in Schedule XII.

SCHEDULE I.

On the procedure to be observed in setting apart Reserved Forests.

1. Whenever a Forest Officer proposes to reserve a forest tract, he shall personally go round and through it to ascertain its natural boundaries, and hold a valuation survey of it.

2. A report containing full particulars regarding the tract, whether any inhabitants are included within it, and the reasons for reservation, together with a map of the district showing the exact situation of the tract in question (a sketch map two miles to the inch, if no district map exists), shall be forwarded to the Deputy Commissioner of the district, who will attach his opinion to the report, and forward it through the Commissioner to the Conservator of Forests, who will submit it to the Chief Commissioner for his final orders, after attaching any remarks he may have to make.

3. The Deputy Commissioner, should he see no objection to such reservation, will, on receipt of the report from the Forest Officer, issue immediate instructions to prevent the cutting of gardens and toungyas within the locality which it is proposed to reserve, pending the final orders of the Chief Commissioner.

4. The Chief Commissioner will pass orders on the report, and, if he approve of the tract being reserved, the Forest Officer will proceed finally to demarcate it in accordance with Rule XIV.

SCHEDULE II.

Streams which are to be kept open under Rule XVII.

Rangoon District.
The Pegu or Zamayee river down to Pegu town, and all its tributaries.
Pounglin and all its tributaries.
Mayzel and all tributaries.
Thayet   ditto.
Thabew   ditto.
Magayee   ditto.
Okkan   ditto.
Thounzai   ditto.

Myanoung District.
Beeling and all tributaries.
Wayhla   ditto.
Tanapo   ditto.
Mokkha   ditto.
Minhla   ditto.
Tsapoke and all tributaries.
Tyemyoke   ditto.
Myoung   ditto.
Kadet   ditto.
Gamoong   ditto.
Toung Khyoung ditto.
Boben   ditto.
Nimboo   ditto.
Tounyoh   ditto.
Shway-lay   ditto.
Meimakan or Hline river from Tsanyouay up to Engmah.

Prome District.
Nawing and all its tributaries.
Boolay, including the Choung Goungyee, the Pudday, and all their tributaries.
Keenee and all its tributaries.
Partly Rangoon District and partly Tounghoo District.
Kawleyah and all its tributaries.
Binedan   ditto.
Pyoontazah and
Yainway   ditto.

Tounghoo District.
Koon and all its tributaries.
Phyoo   ditto.
Nagagyat   ditto.
Kwaymathay   ditto.
Khaboung   ditto.
Swah and all its tributaries.
Myolah   ditto.
Bimbyay   ditto.
Gwaythay   ditto.
Koonoog or
Swoaydoyh   ditto.
Kannee   ditto.
Thoukyaghat   ditto.
Youkthawah   ditto.
Moong   ditto.
Padah   ditto.
The Sittang river above Tounghoo town.

SCHEDULE III.

Names of rivers on which the use of Marking-Hammers is prohibited within the British Territory.

The Salween river.

The Sittang river.

The Pegu river at and below Pegu town.

The Puzzoondoung river below Kyoongalay village.

The Meimakah, Hline, and Rangoon rivers.

The Irrawaddee river and its branches, including the Bassein river.

SCHEDULE IV.

Exhibiting the Rules under which Government timber in the Province of British Burma may be disposed of.

1. By public auction sales at stations or in the forests. At all public sales one portion of the payment (not less than 20 per cent. on the amount) is to be made on the day of sale, either in cash, public securities, or promissory notes, and the balance within a term not exceeding three months after the day of sale. No timber to be delivered before payment in full shall have been received, but the timber to be at the risk of the purchaser from the moment it is knocked down.

2. By private sales on indent approved of by the Conservator of Forests, or the chief Forest Officer at the station, or in the division where the sale is held.

Private sales shall not be effected at the rates below the average rates realized at the public sales of the preceding year, unless by special order from the Conservator of Forests.

No private sale of Government timber is to be considered as concluded, and no timber is to be delivered on account of the same, unless the full amount of the purchase-money is paid to the Forest Officer conducting the sale.

3. By sale of the seasoned timber standing or lying in a certain forest district, to permit or lease-holders. This timber remains the property of Government until the full amount of purchase-money is adjusted. These sales hold good for a certain period. Timber not brought out within the time fixed, whether paid for or not, remains the property of Government. Sales of timber under this rule are not ordinarily to be concluded for a longer period than three years.

4. By grant gratuitously to parties residing in the district near the forests, and at a distance from the principal rivers, for the erection of buildings for the common benefit of the public, as Christian churches or chapels, schools, kyoungs, zayats, bridges, &c. Applications for the timber under this rule may be made to the Conservator of Forests, or his Assistants. Final orders regarding such applications granting timber for these purposes, or otherwise, will be issued by the Conservator of Forests, or by such Assistants as he may specially empower so to do. Timber granted under this rule will be pointed out to the grantees by the forest goungways. The grantees must fell and remove the same to the place where the timber is to be used within a fixed time.

Timber for these purposes will be given gratuitously, but on a certain date, which will be mentioned on the grant, this document must be returned to the Assistant or Deputy Conservator of Forests in charge of the division, with an account showing how it has been used. Should it not have been expended for the purposes specified in the application, the timber will revert to Government; and the parties who received the grant will, if they should have applied the timber to other purposes, be liable to punishment under the forest rules.

SCHEDULE V.

Form of Pass for Timber sold by the Forest Department.

No.	Date.	Owner or Consignee.	Place of Destination.	Description of Timber.	Timber marks	No. of logs or pieces.	Total.	Remarks.

SCHEDULE VI.

Form of Pass for Foreign Timber imported into the British Territory.

No.	Date.	Owner or Consignee.	Place of Destination.	Description of Timber.	Timber marks.	No. of logs or pieces.	Total.	Amount paid on account of Duty.	Amount still to be paid.	Where to be paid.

SCHEDULE VII.

Rules for the management of the Kuddo Revenue Station.

1. All timber passing down the Salween and Attaran rivers must be reported at the Kuddo revenue station, within seven days of its arrival, failing which, on discovery, it will be liable to pay double duty before the timber is cleared.

All timber to be reported at Kuddo within seven days.

2. Attaran timber to be declared for duty at the Nantay station where it will be passed, all other timber to be declared for duty at Kuddo.

Attaran timber declared for duty at Nantay.

3. The report must contain full particulars regarding the timber, its position at the station, number and description of logs, their marks, and other particulars. Owners of timber or their agents who give incomplete or incorrect reports, render themselves liable to punishment under the rules for the administration of forests.

Report to contain full particulars.

4. The reports will be open to public inspection for seven days, before a certificate of registry will be granted. Copies of the reports will be forwarded daily to the Moulmein office for public inspection.

Report open for inspection for seven days.

5. After the lapse of seven days, the officer in charge of the revenue station will grant a certificate, and will mark each log with such marks as may from time to time be directed, except in the case of an incorrect report, when a fresh report will have to be made under Rule II., before certificate of registry will be granted. No timber will be passed without a certificate.

Granting of certificate.

6. The following rates of duty are to be paid before the timber can be allowed to leave the station :—

Rates of duty to be paid.

	Rs.	A.	P.	
For logs of and above 5 feet in girth	2	12	0	per log.
For logs of and from Attaran	4	0	0	„
For logs below 5 feet in girth	1	6	0	„
For logs from Attaran	2	0	0	„
Stem pieces	0	9	0	„
Ship crooks	0	4	0	„
Boat „	0	1	0	„
Small „	0	0	6	„
„ pieces	0	2	0	„

On the payment of these rates the timber will be passed and marked with the pass hammer of the timber revenue station. In measuring timber the middle girth must be taken.

Timber passing limits of stations unmarked, liable to confiscation.

7. All timber which passes the limit of the revenue stations without the pass marks of the Forest Department, is liable to be seized and confiscated.

Two years allowed for payment of duty, and removal of timber.

8. All duty must be paid and timber removed within two years of date of certificate, failing which, it will be advertised and sold to recover the duty and clear the depôt.

9. Timber which has been passed must be removed within one month, or be subjected to a fine of eight annas per log per day during one month, after which, if not removed, notice will be given by advertisement, and the timber removed from the limits of the station, at the risk and expense of the owner. If from any natural cause beyond the control of the owner, it is impossible to remove the timber within the period named, application should be made to the revenue officer who will pass special orders in the case.

Time allowed for passed timber to remain at depôt.

10. The time for the removal of timber from the station is from sun-rise to sun-set, and no timber must be removed at any other time without special permission from the revenue officer, who will judge of the necessity of the case. Breach of this rule will be punished under the forest rules.

Timber when to be removed.

Timber from Government forests—how to be passed. 11. Timber from the Government forests will be passed by the timber revenue officer upon a certificate granted by the Forest Officer showing that the forest revenue has been adjusted.

Agents—how appointed. 12. Parties wishing to transact business through agents must duly appoint these agents as required by law. The names of agents must be registered at the revenue office.

Obliterations of marks—how punished. 13. All timber on which marks have been purposely obliterated by fire, or otherwise obliterated or defaced, will be retained by the Kuddo revenue department, and be treated as drift timber under Rules 27 and 28 of this Schedule.

Lien on timber to be held by Government. 14. Government to have a lien on all timber, whether decreed by the civil courts or not, until payment of all Government demands on the timber has been made.

Channel where timber is not to be placed. 15. Timber brought to the revenue station is not to be placed in any part of the channel on the east side of the Koutsing Island. Any timber found lying there will be removed at the risk and expense of the owner.

River frontage at depôt—how obtained. 16. Parties desiring to have river frontage allotted to them at Kuddo for placing and collecting their timber, must apply to the revenue officer who will pass the necessary orders. The portion allotted will be registered in the office, and the application must be renewed on the 1st of January of every year.

Marks and hammers to be registered. 17. Timber owners wishing to collect and mark their timber at the Kyodan, must register their marks and hammers at Kuddo.

Certificate to contain all marks put upon timber. 18. After the receipt of certificate of registration, the timber may be stamped by purchasers at Kuddo, but the marks or stamps put upon it must be reported at the office either in Moulmein or Kuddo, to be entered in the original certificate which should bear a complete record of all marks on the log entered in it.

Fee for registration of hammers. 19. A fee of ten rupees must be paid for every mark or hammer registered at the Kuddo revenue office, and a certificate will be granted on payment of fee.

Registration to extend over two years. 20. This registration to hold good for two years, that is to say, from 1st May in each year to 1st May of the second year, when parties are required to register their hammers, either in person or by duly appointed agent.

Breaches of rules—how punished. 21. Breaches of these rules by subordinates of the Forest Department and others, will be punishable under the rules for the administration of the forests in British Burma.

Appeal—where to lie. 22. Appeals from the decisions of a Kuddo revenue officer shall lie with the Deputy Conservator in charge of the Salween division, and from the latter with the Conservator. Appeals from the decisions of the Conservator of Forests shall lie with the Chief Commissioner.

Appeals in special cases—where to lie. 23. In the absence of the Conservator or Deputy Conservator from Rangoon or Moulmein, an appeal from the orders of the Kuddo revenue station officer shall lie with the Deputy Commissioner of the district of Amherst, who will pass orders, should he consider it necessary to do so, after enquiry into the matter. Such orders may be appealed from, to the Commissioner, Tenasserim, whose order shall be final.

Drift timber below Kuddo. 24. Parties who have salved timber below Kuddo must bring it to one of the Government drift timber depôts, reporting their having done so, with full particulars, at the forest office, Moulmein, where the amount of salvage, as per Schedule IX, will be paid to them on the timber being taken charge of.

Drift timber above Kuddo. 25. Parties who have salved timber above Kuddo must bring it to one of the stations mentioned in Schedule IX, where they will be entitled to receive the salvage due.

Salvors to be appointed for salving of drift timber. 26. Salvage of timber below Kuddo will be in the hands of salvors (not to exceed four in number), to be named by the trade and approved of by the Forest Department, who will issue a "letmhat" to each, and register their names in the forest office. These "letmhat" will be cancelled by the Forest Department on sufficient grounds being shown.

Ownership to drift timber arriving at Moulmein—how to be proved. 27. All drift and unclaimed timber coming down the Salween, Beeling, Attaran, and other rivers to Moulmein, will be considered the property of Government, unless proof of ownership be given within the time provided in Rule 28.

Regarding time allowed for claim. 28. Monthly notices will be issued of the number of drift logs brought in during the month. Lists containing the description of the logs and their marks (so far as they have been ascertained), will be open for inspection at the Kuddo and Moulmein offices. Claims to drift timber will be admitted up to date of preliminary notice of sale.

29. Public sales of drift timber (should sufficient have been collected to render a sale advisable) will take place every six months, on the 15th January and 15th July in each year. At these sales no timber will be sold, the receipt of which has not been publicly notified, inviting claimants to come forward under Rule 28, six months previously.

Regarding public sales of drift timber.

30. Logs found beached or aground below the mouth of the Attaran, and not in charge of any person, will be considered to be drift timber and treated as such.

Logs found beached to be treated as drift timber.

SCHEDULE VIII.

Names of River Stations where all Timber will have to stop to be examined.

RANGOON STATION.

Timber arriving from the Sittang river.	At the mouth of the Poozoundoung creek.
Timber arriving from the Irrawaddee river.	At the mouth of the Taan Khyoung below Kemendine.

BASSEIN STATION.

Timber arriving at Bassein must stop above the town at such a place as the Collector of Timber Revenue may direct.

SCHEDULE IX.

Rates of Salvage for Timber saved from the undermentioned Rivers :—

	Rs.	A.
I.—Sittang river station above and including Kyassoo ... per log	3	8
If brought to Rangoon ... ... „	6	0
II.—Irrawaddee river stations above and including Henzadah ... „	2	0
If brought to Rangoon or Bassein ... ... „	4	0
III.—Hlyne, Pegu, and Pounglin rivers, also Rangoon river from and above Monkey Point, delivered at Government depôt ... ... „	2	0

Below Monkey Point and lying on sea-shore, delivered at Government Depôt, expenses incurred to be paid.

One-half only of the rates set down will be paid for drift timber saved, but not brought into Government stations and made over to Forest Officers.

SALWEEN RIVER.

Below Kuddo.

	Rs.	A.
Between Kuddo and Battery Point depôt ... ... per log	3	0
„ Battery Point depôt and Kulmee channel ... „	7	0

Above Kuddo.

						Rs.	A.
At Kyodan	...	...	...	...	„	0	4
„ Kamanlay	...	...	...	...	„	0	8
„ Koman or Komoung	...	...	...	...	„	1	0
„ Meebong to Palian	...	...	...	...	„	2	0
„ Kuddo	...	...	...	...	„	3	0

For logs in length below 12 cubits, or in girth below 2 cubits, and for logs that are burnt or otherwise of inferior value, reduced rates will be paid.

SCHEDULE X.

Form of Permit to collect Drift Timber.

_____ is permitted to pay salvage on, and collect the logs which bear the marks entered below in this Permit.

Timber marks.

KUDDO_____

_____186 .

N. B.—This Permit is good only up to the 1st January 186

For salvage rates, see back of Permit.

SCHEDULE XI.

Form in which claim for Drift Timber must be prepared.

_____ claims _____ logs of drift timber lying at _____

as per description and marks entered below :—

Depôt No.	Description and number of logs.	Marks.	Orders of Revenue Officer and Remarks.

SCHEDULE XII.

Register of Forest Cases tried and decided by Forest Officers.   186 .

Number of Register.	Name, residence, and occupation of the accused.	Date and locality when and where the offence was committed.	Nature and description of offence.	Nature of evidence.	Date of sentence.	Sentence.	Commutation of punishment.	Remarks.

B.—ACT No. VII of 1869.

_____

PASSED BY THE GOVERNOR GENERAL OF INDIA IN COUNCIL.

(Received the assent of the Governor General on the 12th March 1869.)

AN ACT TO GIVE VALIDITY TO CERTAIN RULES RELATING TO FORESTS IN BRITISH BURMA.

WHEREAS certain rules for the better management and preservation of the Government forests in British Burma, dated the 2nd day of August 1865, were framed under Act No. VII of 1865 (*to give effect to rules for the management and preservation of Government Forests*), and were confirmed by the Governor General of India in Council and published in the *Gazette of India* dated the 12th day of August 1865 ; and whereas certain of the said rules relate to timber not the produce of such forests, and it is expedient to validate such rules and to indemnify the officers and other persons who have acted under them ; It is hereby enacted as follows :—

Preamble.

1. The rules for the better management and preservation of the Government forests in British Burma, dated the 2nd day of August 1865, shall, from such day down to the passing of this Act, be deemed to have had the force of law as regards all timber to which they purport to relate, and shall continue in force until the said Governor General in Council shall otherwise order.

Validation of Burma forest rules.

2. All officers and other persons are hereby indemnified for anything done before the passing of this Act which might lawfully have been done if this Act had been in force ; and no suit or other proceeding shall be maintained against any such officer or other person in respect of anything so done.

Indemnification of officers.

C.—ACT No. XIII of 1873.

PASSED BY THE GOVERNOR GENERAL OF INDIA IN COUNCIL.

(Received the assent of the Governor General on the 7th August 1873.)

AN ACT TO AMEND THE LAW RELATING TO TIMBER FLOATED DOWN THE RIVERS OF BRITISH BURMA.

Preamble. WHEREAS it is expedient to amend the law relating to timber floated down the rivers of British Burma; It is hereby enacted as follows:—

I.—*Preliminary.*

Short title. 1. This Act may be called "The Burma Timber Act, 1873":

Local extent. It extends only to the territories for the time being under the administration of the Chief Commissioner of British Burma;

Commencement. And it shall come into force on the 7th day of September 1873.

Enactments repealed. 2. The enactments mentioned in the schedule hereto annexed are repealed to the extent mentioned in the third column of the said schedule.

II.—*Duties.*

Duty on timber. 3. On all timber the produce of forests situate beyond the frontier of British Burma and floated down any of the rivers of the Martaban and Tenasserim Provinces, or past a frontier Custom House on the Irrawaddy or Sittang, a duty shall be levied, in such manner, at such places, by such persons, and to such extent, as the Chief Commissioner, with the previous sanction of the Governor General in Council, from time to time prescribes by notification in the *British Burma Gazette :*

Provided that, if the duty is directed to be levied on each log of the said timber, it shall not exceed two rupees twelve annas per log of five feet in girth and upwards, and one rupee six annas per log of less than five feet in girth :

Provided also that, if the duty is directed to be levied *ad valorem*, it shall not exceed eight per cent. on the value.

Power to fix value for assessment of duty. 4. In every case in which such duty is directed to be levied *ad valorem*, the Chief Commissioner may, with the like sanction, from time to time fix by like notification, the value on which the duty shall be assessed.

Legalisation of prior levy of duty. 5. All duties levied before this Act comes into force on timber floated down any of the rivers of British Burma, shall be deemed to have been levied in accordance with law.

Indemnity. And all officers and other persons are hereby indemnified for anything done before this Act comes into force, which might lawfully have been done if this Act had been in force; and no suit or other proceeding shall be maintained against any such officer or other person in respect of anything so done.

III.—*Rules.*

Rules for timber floated. 6. The Chief Commissioner may from time to time, with the previous sanction of the Governor General in Council, make rules as to all or any of the following matters :—

(*a*)—the time and manner of floating timber and bamboos down any of the rivers of British Burma,

(*b*)—the detention and examination of timber and bamboos so floated,

(*c*)—the marking of timber so floated, and the use and registration of timber-marks and marking-hammers,

(*d*)—the sale of unclaimed timber so floated,

(*e*)—the salving and collecting of timber so floated,

(*f*)—the protection of timber in transit,

(*g*)—the descriptions of timber which may be lawfully floated, and the descriptions of timber which may not lawfully be floated down any such river,

(*h*)—the collection of royalty due on timber or bamboos so floated, and·

(*i*)—such other matters connected with timber and bamboos floated down any of the rivers of British Burma as the Chief Commissioner may from time to time think fit to regulate.

And all timber and bamboos found on, in or by any of such rivers, or stranded on the seashore, or adrift on the sea, which has been floated contrary to any rules made under this section and for the time being in force, may be confiscated.

SCHEDULE.

(See Section 2.)

Number and year.	Title.	Extent of repeal.
XXX of 1854.	An Act to provide for the levy of Duties of Customs in the Arracan, Pegu, Martaban and Tenasserim Provinces.	Sections one, seven, eight and twelve.
		Sections four and six, so far as they relate to timber floated down the rivers of British Burma.
IV of 1863.	An Act to give effect to certain Provisions of a Treaty between His Excellency the Earl of Elgin and Kincardine, Viceroy and Governor General of India, and His Majesty the King of Burma.	The whole Act, so far as it relates to duty on such timber.
VII of 1869.	An Act to give validity to certain rules relating to forests in British Burma.	The whole Act, so far as it gives validity to rules relating to duty on such timber.
	The Rules for the better management and preservation of the Government forests of British Burma, dated the 2nd August 1865.	So far as they relate to duty on such timber.

* D.—DRAFT RULES FOR THE CONTROL OF RIVERS, THE TRANSIT OF TIMBER, AND THE MANAGEMENT OF TIMBER DEPOTS AND REVENUE STATIONS IN BRITISH BURMA.

THE following rules having been made by the Chief Commissioner of British Burma in the exercise of the power conferred on him by the Burma Timber Act, 1873, with the previous sanction of the Governor General in Council, are hereby promulgated for general information :—

CHAPTER I.

Of the protection of rivers used for timber floating and of timber in transit.

THE control of all rivers, streams, creeks, and shores in British Burma, as regards the floating of timber and bamboos, and of all stations and places thereon at which any royalty, toll or duty on timber or bamboos is levied, or which are used for the detention, examination or storing of timber or bamboos, is vested in the officers of the Government Forest Department : *Provided* that where any such river, stream, creek or shore, or place or station as aforesaid, is not within the limits of any forest division, the Chief Commissioner may appoint any person to exercise all or any of the powers of a Forest Officer for the purposes of the Burma Timber Act, 1873.

[margin: Control vested in officers of Forest Department.]
[margin: Proviso.]

2. In these rules, unless anything in the context denotes the contrary, "timber" means timber floated down on any of the rivers of British Burma, and includes "bamboos." "Forest Officer" means any officer of the Government Forest Department, and includes a person appointed under Rule 1. "Officer in charge of a forest division" includes a person appointed under Rule 1 to exercise the functions of an officer of the Government Forest Department in charge of a division.

3. The closing or blocking up either partially or entirely, for any purpose, without the joint permission of the officer of the civil district and of the officer in charge of the forest division, of the rivers and streams set forth in Schedule I annexed to these rules, is prohibited. It shall be lawful for the Chief Commissioner from time to time to add other rivers and streams to the list by notification in the official *Gazette.*

[margin: Closing of streams prohibited.]

The throwing of toungya, orchard or garden refuse into these rivers is prohibited.

4. No timber which, under these rules, is subject to examination, or which has to pay any duty or royalty, or requires to be stamped or marked before passing out of the control of the Forest Officer, or which is stranded, sunk or adrift in any river, stream, creek or on any shore of British Burma, shall be marked, nor have any mark on it effaced or altered, nor shall it be converted, sawn, chipped, split, squared, hollowed, cut into pieces, burnt, concealed, removed, sold, or otherwise disposed of, until it has been lawfully examined, paid for, passed, and marked, as the case may be, or until the special permission in writing of the Conservator of Forests, or the officer in charge of the forest division, has been obtained.

[margin: Defacement and conversion of timber prohibited.]

5. No person shall, without the permission in writing of the Conservator of Forests, or officer in charge of the forest division, carry, use, or have in his possession, any marking-hammer or other

[margin: Use of marking tools on certain rivers.]

---

* These Rules are printed as revised by Mr. Baden-Powell. The Memorandum of Objects and Reasons was written prior to the revision, hence the numbering of the Rules does not in all cases agree with the numbers entered in the Memorandum.

instrument used for the marking of timber on any of the rivers, creeks or shores following, that is to say,—

    I.—Salween river and its tributaries.
    II.—The Beeling and Doomdamee rivers and the Kyouktsareet creek.
    III.—The Hlynebwé, Houndraw, Zammee, and Winyeo rivers.
    IV.—The Sittang river and the Kyatsoo creek.
    V.—The Pegu river at and below Pegu town.
    VI.—The Puzoondoung or Pounglin river below Kyoongalay village.
    VII.—The Myetmakba, Hline or Rangoon rivers.
    VIII.—The Irrawaddy and its branches, including the Bassein river.
    IX.—The mouths of the above rivers where they discharge themselves into the sea.
    X.—The sea-shores of British Burma.

It shall be lawful for the Chief Commissioner, from time to time, to add rivers and creeks to this list by notification in the official *Gazette.*

Special rules regarding marking tools and marks on the Salween. 6. On the Salween river the following rules shall be enforced :—

    (a).—All importers of, or traders in timber, and all other persons bringing timber down the Salween river or its tributaries, from Government forests or from foreign territory, shall register at the Kadoe forest office the mark used by them as a timber property-mark. No mark which has not been duly registered shall be recognised by any Forest Officer.

    (b).—The officer in charge of the forest division may refuse to register any timber-mark or marks, when registration is applied for by or on behalf of a minor or a married woman.

    (c).—On and after the date on which these rules come into force, all persons applying for registration for the first time, or applying to register a mark which has once been registered but not continuously maintained on the register, shall be permitted to register only one mark for each person : *Provided* that any such person may, on sufficient cause being shown, register one mark as his original and principal mark, and one or more marks as additional thereto.

Such additional mark or marks shall in all cases consist of some combination into which the original mark, unaltered in itself, enters.

All other persons shall be entitled to register as many marks as they have hitherto kept continuously on the register.

Illustration.—A person applying for the first time, or having once in 1864 registered the mark, but not kept it up, now registers B as his principal mark, he applies to register additional marks on the ground that he has several forest working divisions, the timber from which he desires to distinguish for purposes of account. He may be permitted to register such further marks as B, B2, B x

    (d).—A fee of Rs. 100 shall be paid for every principal mark registered under rule (c), and additional marks, when registration thereof is permitted, shall be registered without additional fee. A fee of Rs. 10 shall be paid for registration in all other cases.

    (e).—Every registration shall hold good for three years; that is to say, from the 1st April in any year till the 31st March of the third year following. On the fee being paid and registration effected, the officer in charge of the forest division, or of the Kadoe sub-division, shall grant a certificate under his signature and the seal of the office, showing that mark or marks have been registered and the fee paid.

CHAPTER II.

Of timber adrift, unclaimed, defaced, or left without control.

7. All unclaimed timber adrift on any of the rivers, creeks, or off the sea-coast of British Burma, and all timber on which the marks have been obliterated, altered or defaced by fire or otherwise, and all timber beached, stranded or sunk in the rivers, streams or waters, or on the sea-shore of British Burma, may be collected by any Forest Officer, and shall in all cases be taken to one of the stations enumerated in Schedule II annexed to these rules, and there detained till its disposal under these rules is determined.

Drift timber *prima facie* the property of Government.

Its treatment.

Process of treatment not to be interfered with.

8. At the stations of Rangoon, Prome, Toungoo, and Moulmein, and at such other stations as the Conservator of Forests may direct, notices shall be published on the last day of every month stating the number and description of pieces of drift timber brought in during the month, and requiring claims to the ownership of such timber to be lodged within three months from the date on which the notice is issued.

Notices of drift timber to be published.

Proviso.

Provided that such notices shall not be issued in the case of drift timber carried down to the sea-shore beyond control ; but the Conservator of Forests may direct the local sale of such pieces for the benefit of Government. Nothing in this proviso shall be held to prevent the Conservator of Forests, or officer in charge of the forest division, from granting a written permission to any owner who can show to his satisfaction that he is able to identify such timber and remove the same.

9. All claims to timber notified under the last preceding rule shall be submitted personally or by duly authorized agent, with full particulars, and in the form prescribed in Schedule III annexed to these rules. If within three months as aforesaid no claimant appears, or having appeared, fails to establish his title, the timber shall be sold on account of Government.

Claims to timber; how submitted.

10. All claims to drift timber shall be decided by the Conservator of Forests or by the officer in charge of the forest division. From all such decisions when the timber exceeds Rs. 100 in value, an appeal shall lie to the Commissioner of the division: *Provided* that in any case when two or more persons are claimants to the same timber the Conservator of Forests or Forest Officer aforesaid may decline to decide the case and refer the parties to the civil court. The Chief Commissioner may prescribe the procedure to be followed in hearing and deciding such claims and appeals.

Claims how decided.

Appeal.

Procedure.

11. Timber awarded to claimants shall not be delivered until redeemed by payment of all salvage money, duty, royalty or other charges which may have been incurred, or may be due on account thereof.

Timber awarded to be redeemed.

12. Public sales of drift timber to which no claim has been made or proved under Rule 9 (should a sufficient quantity have been collected to make a sale advisable) shall take place on the 15th January and 15th July in each year, unless the Conservator of Forests shall otherwise publicly notify.

Sales of waif.

CHAPTER III.

Of the collecting and salving of timber.

13. Owners who wish to collect their own timber which has not yet been formed into drafts or which from any cause is floating loose and without control, shall, on the rivers, streams and shores named below, collect such timber by virtue of a "collecting license" issued by and at the discretion of the officer in charge of the forest division or sub-division and not otherwise. Such license shall specify the kind of timber to be collected, the limits within which collection of such timber is to take place, and the whole of the marks which the timber bears; it shall continue in force for one year, and shall be returned to the office whence it was issued, by the 31st December in each year.

Collecting timber by the owners thereof.

Every person collecting timber on behalf of an owner under this rule shall be severally furnished with a license, and no person shall be entitled to collect timber in virtue of a license in the hands of some other person collecting on behalf of the same owner.

Each individual to have a license.

The Forest Officer as aforesaid may refuse to issue a collecting license for any timber, the ownership of which is in dispute, or is the subject of litigation, and may, in lieu of such license, issue an order in writing directing such timber to be collected and brought to one of the drift timber depôts specified in Schedule II annexed to these rules.

Refusal to issue license.

The rivers, streams and shores to which this rule is applicable are the following :—

Rivers on which the rule is in force.

    (1.)—The Salween river and its tributaries.
    (2.)—The Beeling and Doomdamee rivers and Kyoukstareet creek.
    (3.)—The Hlynebw, Houndrawé, Zammee, and Winyeo rivers.
    (4.)—The Sittang river and the Kyatsoo creek.
    (5.)—The Pegu river at and below Pegu town.
    (6.)—The Puzoondoung or Pounglin river below Kyeongalay.
    (7.)—The Myetmakha, Hline or Rangoon rivers.
    (8.)—The Irrawaddy river and its branches, including the Bassein river.
    (9.)—The mouths of the above rivers where they discharge themselves into the sea.
    (10.)—The sea-shore of British Burma.

Collecting licenses on the Salween shall not hold good below the Martaban Forest Department guard-house, but all collection of timber below that point shall be managed as hereinafter provided.

14. On the Salween river, at the season of the year when the collecting of timber is effected by the aid of ropes stretched across the river, persons having collecting licenses under Rule 13 shall collect their timber under the superintendence of the officer in charge of the forest division or other officer appointed by him in that behalf.

Management of the rope station on the Salween.

On such timber being collected and formed into rafts, the said Forest Officer shall inspect such rafts at Yembaing on the Salween, and shall issue a pass for all rafts leaving that place; no raft shall be allowed to proceed down the river unless covered by such a pass. Every such pass shall

Pass for rafts at Yembaing.

specify the number of logs, the date of despatch, the name of the person claiming to be the owner, the number and names of the raftsmen, and shall be signed by the Forest Officer aforesaid.

Every such pass shall be issued uniformly from a book furnished with a counterfoil on which a note shall be made under the signature of the Forest Officer of all the entries in the pass.

15. Except as provided for the Salween river in Rule 16, the *bonâ fide* salving of timber, *Bonâ fide salving of timber unrestricted.* that is to say, the securing of timber by any person not being the owner which has gone adrift, so as to preserve it from danger or from being carried away beyond control, is unrestricted.

Timber so salved shall (subject to the provisions of Rule 17 *post*) be made over to *Salved timber how to be disposed of.* such persons as have licenses under Rule 13 to collect it.

When timber so salved is not claimed by any person having license or authorized to collect it, or bears no mark, or has had the marks altered or defaced in any way or is unidentifiable, the salvor shall make it over to any Forest Officer collecting timber under Rule 7, or shall retain it at the bank or place of salving until the Forest Officer, as aforesaid, comes to take it away. Timber salved by boats and vessels in the course of their voyage shall be taken to the nearest drift timber or other forest station, and made over to the Forest Officer in local charge.

Salving rules on the Salween. 16. On the Salween river the following special rules shall be in force regarding the salving of timber :—

(a).—On the Salween above the mouth of the Gyochoung, salving shall be effected only by persons having "salving licenses" from the officer in charge of the forest division. Each salving license shall prescribe the limits within which salving is to take place and shall bear a serial number in English and Burmese, and every license-holder shall be furnished, at his own expense, with a marking-hammer bearing his license number.

Such licenses and hammers shall hold good and be used for one year only, and shall be returned to the office whence they were issued by the 31st December in each year.

(b).—All persons having "salving licenses" shall exhibit the number of the license on their huts, rafts and boats in English and Burmese numerals of a conspicuous size ; and all logs salved shall be marked with the hammer of the licensed salvor.

(c).—Licensed salvors shall make over the salved timber to persons duly authorized to collect, according to Rule 13, and such salved timber as may not be applied for by any one duly authorized to collect timber as aforesaid, may be kept, and at the close of the rainy season be brought down to the Kadoe timber revenue station, or made over to a Forest Officer collecting timber under Rule 7.

Timber so brought down or made over shall be dealt with as drift timber, according to the provisions of rules 5 to 12, both inclusive.

(d).—Below the mouth of the Gyochoung and down to the Forest Department guard-house at Martaban, *bonâ fide* salving is unrestricted.

(e).—Below the Forest Department guard-house at Martaban, officers shall be appointed on such salary and on such terms as the Chief Commissioner may direct, to be called "Government Timber Collectors." Any person may salve any timber in such portion of the river, but he shall forthwith make it over to the Government Collector at the Battery Point timber depôt, or report the salving thereof at the forest office at Kadoe or Moulmein, and retain the timber until its disposal is ordered by the Government Timber Collector.

17. All persons who have lawfully salved timber shall be entitled to recover salvage money *Salvage dues.* at the rates prescribed in Schedule IV annexed to these rules : and no such person shall be bound to give up the salved timber until salvage payment thereof at the prescribed rate is tendered.

It shall be lawful for the Chief Commissioner to alter the rates prescribed in the said Schedule IV from time to time : and a notification setting forth the salvage rates and alterations therein, shall from time to time be published for general information.

18. In any case when salving is restricted by these rules, timber salved by unauthorized *Salving by unauthorised persons.* persons shall, nevertheless, be given up to any person duly authorized under Rule 13 to collect it, or it may be taken charge of by any Forest Officer collecting timber under Rule 7, and may be dealt with as drift timber.

No unauthorized person shall be entitled to any salvage or other remuneration in respect of any such timber.

CHAPTER IV.

Of the stoppage, examination and passing of timber.

19. All foreign timber when brought across the British frontier into the Thayetmyo or *Report of timber in transit on certain rivers.* Toungoo districts, and all other timber floated on the Irrawaddy, Bassein, Hline, Pounglin, Pegu or Sittang rivers shall be reported at the first timber revenue station it reaches to the Forest Officer at such station. The report shall be in writing and shall contain the particulars required by Rule 22.

All timber may be detained by the order of any Forest Officer in charge of the timber revenue station, until it is so reported, and until the said officer is satisfied, by examination of the timber, or otherwise, that it has been correctly reported and has been lawfully obtained, and until all duty, royalty or other charge that may be lawfully payable in respect of such timber has been paid.

Power of detention.

On such officer being satisfied of the correctness of the report and on the duty, royalty or other charge being duly paid as aforesaid, he shall mark the timber with the Government mark, indicating duty-free foreign timber, or with the mark indicating the payment of duty or royalty as the case may be: he shall then grant a pass for the timber in the form prescribed by Schedule V annexed to these rules.

Marking and passing.

Provided always, that in the case of British timber obtained pursuant to a forest permit, the examination shall be made and the pass shall be issued at such station as the permit prescribes.

Proviso.

No person shall remove or attempt to remove any timber from the station, unless covered by a pass; nor, without permission of the officer in charge of the station, at any time between sun-set and sun-rise.

20. All such timber shall be further subject to stoppage and examination at the stations specified in Schedule VI. No raft shall leave or pass such stations without the order of the officer in charge of the station. Such order shall ordinarily be endorsed on the pass.

Further examination of timber.

The officer in charge of the forest division may also direct the examination of passes *without* stoppage of rafts at any point during transit.

21. All timber passing down the Salween, Beeling, Doomdamee, Benlyne, Houndraw, and Gyne rivers, whether liable to duty or not, shall be taken to the Kadoe timber revenue station. Timber passing down the Attaran shall be taken to the Nantay timber depôt. All such timber shall be reported as hereinafter provided at the Kadoe station office within seven days. Drift timber redeemed under Rules 10 and 11 at Kadoe is subject also to this rule. All timber not so reported shall if liable to duty under Act XIII of 1873 be chargeable with four times the amount of such duty, and if not liable to duty with a penalty equal to four times the amount which would have been leviable had it been so liable.

Salween timber how dealt with.

Penalty for not reporting at Kadoe.

22. The report prescribed in Rules 19 and 21 shall be in writing, and shall contain full particulars regarding the description of timber, the number of the logs, the whole of the marks, the names of the person or persons claiming to be the owners, the number of the land allotment on which it is deposited if at Kadoe, and such other particulars as the Conservator of Forests may from time to time direct.

Contents of report.

In the case of timber rafted on the Salween from the rope stations at Yembaing according to Rule 14, the report shall be accompanied by the pass prescribed by that rule.

Persons reporting timber, who send in incomplete or incorrect reports, shall be deemed to have committed a breach of this rule.

23. In the case of the Salween, Beeling, Doomdamee, Benlyne, Houndraw, Gyne, and Attaran timber, on the report being sent in, each log or piece of timber shall be entered in a register kept for the purpose at the Kadoe office. The register shall provide a running number for each log or piece of timber entered therein, and shall contain such other particulars as are required by the form prescribed in Schedule VII annexed to these rules.

Register at Kadoe.

So soon as the reports shall have been found correct as regards the number, description, and marks of the timber, and the register has been corrected accordingly (when necessary) each log or piece of timber shall be marked with the register number.

24. Every report filed at Kadoe shall lie open to inspection at the Kadoe office for seven days; copies of each day's reports shall be forwarded to the Moulmein forest office, and there lie open to inspection for a similar period.

Reports at Kadoe to be open to inspection.

Provided that any person interested may apply to the Forest Officer in charge of the division to extend the period of inspection of any report for a further period of seven days. And the said Forest Officer, after hearing the objection, if any, made by the person reporting the timber, may, if he sees sufficient cause, extend the period accordingly.

Extension of period.

25. On the expiry of the prescribed period, if in the meantime no order of a civil court or other competent authority has been served on the officer in charge of the forest division or of the Kadoe sub-division, directing him to hold or otherwise dispose of the timber in question, a "teingzú" or certificate shall be granted to the person in whose name the timber has been reported.

Issue of teingzú.

Provided that no such certificate shall be issued unless the applicant produces the certificate of the registration of his mark or marks under Rule 6. The certificate shall be in the form prescribed in Schedule VIII annexed to these rules, and shall be signed by the officer in charge of the forest division or of the Kadoe sub-division, and shall be sealed with the seal of the forest office.

Proviso.

On the issue of the certificate the timber register at Kadoe shall be signed by the said officer, in the column left for the purpose opposite to the entries of the timber for which the certificate is granted.

Signature to the register.

26. In the case of timber sold, mortgaged, or pledged as collateral security at Kadoe before payment of duty or royalty, the timber may be marked by the purchaser, or mortgagee or pawnee, but such mark shall be reported at the Kadoe office and be entered on the timber register described in Rule 23.

Sale of timber at Kadoe before payment of duty.

27. An extract from the timber register described in Rule 23 may be granted on application to the forest office at Kadoe on payment of such fee as the Chief Commissioner may prescribe.

Extract from the register.

28. When all Government duty, royalty and other charges lawfully leviable have been duly paid, the timber shall be marked by the Forest Officer at Kadoe, or Nantay, as the case may be, with a hammer indicating that all dues have been paid and that the timber is finally passed.

Final passing of timber.

On such mark being duly applied, a permit to remove the timber shall be issued in the form prescribed by Schedule IX annexed to these rules.

No person shall remove, or attempt to remove, any timber from the Kadoe or Nantay timber stations unless covered by such a permit.

The permit to remove shall be given up to the Forest Officer on guard at the exit of the timber station, who shall on production thereof allow the timber to pass out.

29. Persons wishing to transact timber business by means of agents shall register the names of such agents at the forest offices concerned, and shall notify, for entry in the register, the extent of the power and authority of such agents. No agent whose name is not registered shall be recognized as such, and the power and authority of agents shall be deemed to be such as are entered in the register; any written instructions or powers of attorney to the contrary notwithstanding. The officer in charge of the forest division may refuse to register any agent, or any powers of any agent, and require the person applying either to transact his business in person or appoint some person as his agent other than the one refused.

Power of agents.

CHAPTER V.

Of the management of the Kadoe Timber Station.

30. The limits of the land belonging to Government and used for the purpose of the Kadoe timber station shall be demarcated by conspicuous marks.

Demarcation of Kadoe.

No person shall reside, or deposit timber within the demarcated limits without the permission of the officer in charge of the forest division.

31. At the upper and lower limits of the station, Forest Officers shall be placed to be called the "entrance guard" and "exit guard" respectively.

Entrance and exit guards.

It shall be the duty of such guards to count every log or piece of timber that enters or leaves the limits of the station, and to keep a book in which the daily number of logs or pieces so coming in and going out, is duly entered.

32. No timber brought to Kadoe shall be placed in any part of the channel on the east side of the Koutsing Island. It shall be lawful for the officer in charge of the forest division or Kadoe sub-division to remove, at the expense and risk of the owners, any timber that may be so placed.

Channel to be kept clear.

33. Applications for allotments of land with river frontage within the demarcated limits of the Kadoe station shall be made to the officer in charge of the forest division or of the Kadoe sub-division. Such allotments shall be made free of charge and shall be numbered and entered in a register kept for the purpose.

Allotments of land at Kadoe.

Such allotments shall only hold good for one year, but applications may be renewed on the 1st January in each year.

No allotment of land shall be made in such parts of the station as the officer in charge of the forest division or of the Kadoe sub-division desires to keep clear for the location of guard-houses, offices or Government timber.

34. Every person to whom an allotment of land has been made may erect such huts or houses thereon as he pleases, and may remove the materials on ceasing to hold the allotment, but on every allotment of land the registered number shall be exhibited on a painted board in English and Burmese numerals of a conspicuous size. The grazing or keeping of cattle in such allotments is prohibited, provided that the officer in charge may grant permission to keep such reasonable number of cattle as the watchman hereinafter provided to be maintained requires for his own use.

Erection of huts, &c.

35. Every person to whom an allotment of land has been made shall forthwith appoint and thereafter maintain (so long as he holds the allotment) a guard or watchman.

Watchman to be appointed.

Such watchman shall be a person of good character, approved of in writing by the officer in charge of the forest division or of the Kadoe sub-division, and his name shall be entered in the register of land allotments.

It is the duty of such guard or watchman to obey the officers in charge in all lawful matters pertaining to the safety, discipline, and good order of the timber station. Should he fail to do so, the officer in charge of the forest division may require the person holding the allotment to dismiss such guard or watchman on pain of forfeiting the allotment if he refuses so to do.

Watchman's duty.

36. The Government shall not be held responsible in any way for the safe keeping of the timber stored on the allotments, but owners and their watchmen may take any precautions they deem necessary to secure their property.

No responsibility of Government for injury.

Provided that in case of any accident or emergency, or when any private or public property is in danger, it shall be the duty of all persons in Government employ, as well as others at the Kadoe station, to give assistance to the utmost of their power to save such property and prevent loss or injury thereto.

Proviso.

37. Every watchman appointed under Rule 35 shall keep a book, called the "daily arrival and despatch book"; it shall be in the form prescribed in Schedule X annexed to these rules.

Watchman to keep arrival and despatch book.

Every report of timber brought to Kadoe shall be accompanied by an extract from the daily arrival book, signed by the watchman, to serve as a voucher for the actual existence of the timber reported, and to indicate its position at the station.

38. No timber shall be taken away from the station nor be moved or shifted within the limits of the station at any time between the hours of sun-set and sun-rise, without the permission in writing of the officer in charge of the forest division or Kadoe sub-division.

Timber not to be moved at night.

CHAPTER VI.

Of certain duties and powers of Forest Officers under the Burma Timber Act, 1873.

39. Any Forest Officer of a superior grade may in case of necessity exercise all or any of the functions of the grades below.

Exercise of powers.

40. All Forest Officers, wherever they are and on whatever duty or employment, have always the duty of preventing any offence against the Burma Timber Act, 1873, or rules made under it, arresting the offender and protecting the property of the State or of private persons while detained at or in charge of any Government timber depôt or station.

Duty of all Forest Officers generally.

But in case they act under this rule at a place beyond the limits of their local charge, they shall communicate what they have done to the officer in local charge, and the local officer shall take such further steps as the matter in hand may require.

41. Any Forest Officer may make a complaint to the Magistrate, or to the police, as the case may be, or apply for a confiscation under Section 6 of Act XIII of 1873, in any case of the breach of the law occurring in his own local charge. But the officer in charge of the division may direct any other officer subordinate to him to complain or prosecute, or may conduct the case himself.

Power to prosecute.

42. No Forest Officer having any charge or duty under the Burma Timber Act of 1873 shall trade, or engage directly or indirectly, in any agency business, or in any work or employment other than his duties as such Forest Officer, except with the permission in writing of the Conservator of Forests.

CHAPTER VII.

Of Penalties.

43. Any Police or Forest Officer may arrest without warrant any person infringing or reasonably suspected of infringing any of these rules. On such arrest being made, the person arrested shall, as soon as may be, be taken before a Magistrate, who may, if he see reasonable cause, order such person to be detained in custody or admitted to bail.

Power of arrest.

44. Any Forest or Police officer may seize any timber in transit or otherwise in respect of which an offence is suspected to have been committed, and any boats, cattle, elephants, tools, or carts which have been used in obtaining or transporting such timber; and when any timber reported at any forest revenue station appears to the officer in charge to have been cut, moved, floated or obtained in a manner contrary to law, he may seize such timber.

Seizure of unlawful timber.

On such seizure being made, a distinctive mark indicating the same shall be put on the property seized, and the officer shall, as soon as may be, apply for the confiscation of the timber under Section 6 of Act XIII of 1873, to any Magistrate having local jurisdiction at the place of seizure.

45. On such application being made, the Magistrate may summon the owner or the person in possession of such timber, and on his appearance, or in default thereof in his absence, may examine into the cause of the seizure of such timber, and may adjudge the same to be confiscated under

Enquiry in confiscation cases.

Section 6 of Act XIII of 1873. When the confiscation of any timber shall be adjudged, the same shall thereupon belong to and vest in Her Majesty.

46. An appeal shall lie from any such order of confiscation, if the property confiscated is over Rs. 100 in value, to the Commissioner of the division in which the seizure was made.

Appeal.

47. For every breach of these rules, or omission contrary to the provisions thereof, and for every abetment of such breach or omission within the meaning of the Indian Penal Code, and for every wilful neglect or transgression of duty by any Forest Officer, the offender shall be liable on conviction before any Magistrate to a fine not exceeding Rs. 500, and to imprisonment of either description for a term not exceeding three months, or to both.

Penalty for breach of the rules.

Provided that nothing in these rules is intended to prevent the prosecution of any person under the Indian Penal Code, and not under the rules, for any offence he may have committed contrary to the provisions thereof.

48. The rules of the Code of Criminal Procedure shall be applicable to the admission of appeals and the trial of cases and appeals under these rules, and to the trial of confiscation cases and appeals as far as may be practicable.

Procedure.

49. In any case of a conviction for breach of these rules, a sum not exceeding one-half of any fine levied may be awarded by order of the Magistrate to the informer by whose aid the offender was brought to justice.

Reward to informer.

STATEMENT OF OBJECTS AND REASONS.

The powers granted by Section 6, Act XIII of 1873, are so complete that it is hardly necessary to offer any detailed legal explanation of the validity of such rules as those now submitted.

The only questions which can possibly arise are in reference to the penalties. The Act does not specify that penalties for breach of rules may be prescribed by the rules; but penalties would seem a *sine quâ non*, since there can be no rule enforced without its " sanction," and the *confiscation* provided by the last section of the Act is obviously one that would not touch the perpetrators of breaches of many of the most important rules.

Apart from the general power to affix penalties implied by the power to make rules, the wording of the last clause, " that the rules may provide for such other matters, &c., as the Chief Commissioner may direct," seem to settle the question.

The rules further develop the subject of confiscation already provided by the Act, by prescribing the method of seizure, the procedure of confiscation, and by the provision of an appeal from confiscation order,—a subject which has been much disputed in Burma already, and needs an authoritative settlement.

It is also trusted that the sections declaring the lien of Government on all timber for revenue and other dues, which is eminently necessary to avoid disputes, will meet approval; also the provision that the signed and sealed certificate of such claim for revenue, &c., and of the consequent lien, shall be admissible as evidence of such lien. It is highly inconvenient that the Forest Officer should be summoned away from his work, perhaps miles away, to prove personally in court a matter of pure official routine. The written certificate of a chemical examiner on so grave a matter as the analysis of a poisoned body, is admissible in evidence on the ground of special convenience : with how much more reason should the certificate that certain timber dues or revenues are unpaid, and that consequently any disposal of such timber is made subject to the State lien for the dues on the timber, be receivable in evidence.

It is to be borne in mind, *first*, that the Burma forest system deals both with timber which comes from British and from foreign forests; and *secondly*, that the system of timber management on the Salween is different from what it is in other rivers, by reason of the peculiar formation of the river bed above the site of the market at Moulmein, and by reason of certain long-established commercial customs in the timber trade, which it is impossible to ignore. The rules are not only concerned, therefore, with the mere levy of a certain sum of duty or revenue, but with the protection of the timber owners from fraud, the safety of the timber, and generally with giving effect to those long-established customs which are the distinctive features of the timber trade.

The rules in fact provide a system of river police. From this arises the necessity of prescribing and curtailing the powers of *owners* dealing, or professing to deal, with their own timber.

The Forest Act VII of 1865, with that nice and surprising infelicity which always causes it to fail us on the very points where alone legislative interference is called for, seemingly prohibits any interference with owners in marking, &c. These are the very people who, under colour of looking after their own, will, if uncontrolled, supermark, deface, and otherwise act fraudulently with regard to all timber they can lay hands on.

To tell us by a legislative Act that we may (with the ordinary criminal law in full force) prevent people from touching timber which they have no right to, is surely mere surplusage, and hence a better power of regulating timber transit than Act VII provides was required, and has now been conferred by the Burma Timber Act.

L

The details of the rules may now be explained in form of a running comment. It is very necessary to observe that almost every detail of these rules has its special bearing on some feature of the timber system, and alterations, if thought necessary, should only be made with a full knowledge of the conditions involved.

The rules are divided into seven chapters.

CHAPTER I.

Of the protection of rivers used for timber floating, &c.

Rule 1 asserts the general right of the State to control the rivers as regards the floating and transport of timber.

Rule 2 provides that the administration of the timber control system rests with the officers of the Forest Department. The proviso to the rule is rendered necessary by the fact that at Bassein as yet no Forest Officer exists, so that the Collector of Customs has to undertake his duties in regard to timber coming down the Bassein river.

Rule 3 provides that the rivers, &c., used for floating may not be closed, as they would otherwise be liable to be, by dams, fishing weirs, &c.

Rule 4.—It is the key of the system that when timber is once out of the forest and in transit, all power of altering its condition, form, or marks ceases until such time as it shall have undergone all the processes of examination and passing to which both for revenue and police purposes it is subject.

Once let it be touched by any one,—the owner or otherwise,—in any way prohibited by the rule, and of course control is at an end.

Rule 5.—The prohibition of the use or possession of hammers for marking on certain rivers (which do *not include* the places where timber is first launched, and where, of course, marking is necessary), is part and parcel of the system of protection of timber. The rule prohibits possession of a tool which *can* only be had for a fraudulent purpose, just as the Penal Code prohibits possession of instruments or materials for coinage.

Rule 6 provides a special rule regarding marks for the Salween. Here the peculiar manner in which foreign timber is dealt with in territories beyond control makes it desirable to know definitely what property-mark each timber owner has, so that, in all subsequent stages, the collection, registration and ultimate passing of the timber, the registered marks may be looked to as accompanying every application for dealing with such timber. The rule about married women and minors is necessitated by the fact that persons used to get several marks registered under the pretence that timber bearing this or that mark was property of a wife, a child or other relation. In this way the claims of persons having made advances on the security of timber supposed to belong to one person were evaded: of course widows trading on their own account are not prevented by the rule from registering their marks like any one else.

The Rule (C) is an effort to limit the number of marks, and introduce greater simplicity in recognizing timber, and the higher fee will deter mere adventurers who really have no timber from registering marks with a view to try and get such a mark affixed to some timber in an out-of-the-way place, and then bring it down as their own.

It would, however, be harsh to apply the limitation of marks or an increase of fees to those old-established traders who have for years had more than one mark continuously on our registers.

In any new case where more than one mark is allowed, it is stipulated that there shall be one original or principal mark, and that all others are to be mere combinations into which the principal is introduced.

Illustrations after the manner of the modern Acts have been added here and elsewhere in the rules, with a view to elucidation.

CHAPTER II.

Of Drift and Unclaimed Timber.

Rule 7.—When timber is fairly out of the charge or control of any person acting for the owner, or when it ceases to be identifiable, the State steps in to prevent the total loss and waste of material that must otherwise ensue.

The timber once taken charge of must go through a regular process of identification, and *meanwhile* other legal powers and administration, civil or criminal, ought not to interfere with the process. Such interference was, I believe, attempted by a former Recorder of Moulmein, to the manifest detriment of order and system, so it seems right to insert a formal protection against interference in the rules. If no one can or will prove that the timber is his, it becomes legal waif, and is the property of the State.

Rules 8, 9, and 10 describe the process of claim and identification with a proviso, because, where the timber is actually washed out to sea, it would be too expensive to bring it back, and it must be sold *in situ* for what it will fetch; it is only in one case, perhaps, in a hundred (for which the rule provides) in which an owner can practically reclaim and remove the timber.

It seems right to grant an appeal from the order deciding ownership: for the Forest Officer may decide that the timber is unidentifiable, or not proved, and the rejected claimant may consider himself aggrieved by such a decision, especially when the timber is of some value. It is, however, unnecessary to allow such appeal when the value is very small.* The officers deciding such cases will be always men of experience, and may fairly be trusted with so much power. If two or more claimants are both asking for the same timber, the decision may be either that it is not identifiable, as belonging to *either* (in which case the State claims it), or the timber may be clearly enough marked with the marks of both parties, as in the great Hmyne-loon-gyee cases, &c. Here we propose that the Forest Officer should make no attempt to decide (unless the parties wish it), but let the question go before the ordinary courts of law.

Rules 11 and 12 need no remark.

CHAPTER III.

Of Collecting and Salving of Timber.

Rule 13.—A very important distinction has to be drawn between "collecting" and "salving"—the former is the act of appropriation by the owner or his agent; the latter a mere act of saving or securing property that would otherwise be lost, without any design to appropriate it.

No one, owner or not, is to collect without a license, which must be given in such a way as to show exactly what the person is entitled to be considered the owner of, and *every one* collecting must hold a license: a man is not to say, "I am one of so-and-so's people, collecting for him: so-and-so has the license," perhaps ten miles off. It may often happen that timber in dispute has got supermarked, so that logs bear the marks of two or three claimants; to whom is the Forest Officer to give the collecting license? The first comer is no criterion, for both may rush to the officer simultaneously: in such cases it is best that he refuse the license and direct some officer or person, probably agreed on by the parties, to collect the timber and bring it to some definite spot. Meanwhile, of course, the parties will go to court, and will in their own interest procure the necessary prohibitory orders securing the timber during the process of litigation.

These licenses are confined to the rivers in which the timber ought to be in rafts and under control, and after it has got out of those small creeks and forest streams, in which it has of course to be left to the discretion of the owner.

Rule 14 makes a special provision for the peculiar state of things on the Salween. Here the upper part of the river is a torrent; the logs are launched, and nature itself prevents any one touching them when once in the water. But from the site of the police post of Mainzek the water becomes, when the floods subside, manageable; a series of stout ropes of twisted rattan are, by the combined aid of several of the chief Thitgongs or timber traders, stretched across the river. Each party watches the logs as they come up: the party at the first rope stops all such logs as seem to bear their marks, get them to one side, and detain them, at the same time pushing the other logs under the rope, to let them go on. On reaching the second rope the same process is repeated, and so on till the last. The portion of the river where these ropes are up, and which extends usually for about a mile, as far as the Yembaing village, is called "the Kyodan." There are sometimes four, sometimes six, and sometimes as many as eight ropes up at intervals.

The management of this place should be under a good native officer; he simply looks to the formation of all rafts of timber stopped and despatched, and gives a pass for each. This pass is of great importance, as it prevents a common fraud. Formerly, timber rafts used to be sent down by the back or Martaban channel instead of to Kadoe, and thus escape and get taken to the saw-pits: if caught, there was the excuse that the current had taken them into the wrong channel; the pass will prevent this, for if the timber has honestly come from the Kyodan, but got carried on to the Martaban channel, it will nevertheless have a pass.

Rule 15.—Collecting timber being thus provided for, the salving of timber is to be unrestricted; for we do not wish to prevent people from preserving property from destruction. Of course salved timber is either marked and identifiable, or not so; if the former, it has to be made over to a person with a collecting license; if the latter, or if no licensed person claims it, then the salvor is bound to give it to a Forest Officer, or to keep it on the bank till such officer comes for it. This is preferable to allowing it to be taken to a depôt, for under pretence of so doing, it might be conveyed to a saw-pit instead. When timber is picked up by a vessel or boat in transit, it is taken in tow and deposited at the nearest forest station. Extreme vigilance and thorough patrolling by the river officers are required to work this rule, as most of the others.

Rule 16.—If the reader will look back to the place where we left the Salween timber at the rope stations, he will readily perceive that, in spite of all precautions, some timber gets past all the ropes, even where they are up, and when they are *not* up, of course all the timber is floating at large: accordingly, from the place where rafting begins to be possible, down to the mouth of the Gyo-choung, a large number of persons annually collect and make their living by salving and bringing the logs to the bank; this is bank-salvage properly so called. To control these people is the object of the rule: they are always so numerous, that salving by any one else is here quite unnecessary. The licensed collectors of timber are of course on the look-out, and so most of the banked timber finds its way to the owners, and the forest guards, &c., are on the look-out for fraud. The self-interest of the salvors is enough to make them very sharp; they will not give up the timber unless perfectly satisfied that the person demanding it is really the owner, for at the end of the season all such timber as has not been claimed is taken down to Kadoe, and gets the higher rate of salvage paid there by Government. The timber so taken down, being unclaimed, is advertized and treated as drift timber.

Below the Gyo-choung, and down to Martaban, only a few stray logs come, and there is no object in restricting salving below Martaban. Again, the river widens out, and there are islands and side channels—a main one, the "Kalwee" to the west, and the "Salween" to the south; all about here, both salving and collecting are under the control of special officers. The salving is difficult and expensive, hence the schedule rates are high.

Rule 17 needs no remarks; nor 18, which denies the right of salvage compensation to unauthorized persons.

<h3 style="text-align:center">CHAPTER IV.</h3>

Of the Examination of Timber.

Rule 19 requires the stoppage and report of timber. This is with a view to payment of revenue or duty. This business transacted, a pass is granted, and the timber goes on, being moved always in the hours of daylight to prevent fraud.

Rule 21.—The Salween, again, has a special practice of its own. Here all timber goes to the Kadoe timber station, except that of the Attaran, which goes to Nantay; but the report and other process of treatment also goes on at Kadoe. The reports lie open for inspection, so that if an owner has reported timber that some one else claims, he may at once be aware of the fact, and be able to institute proceedings in court. Generally speaking, seven days is enough time; but it is right to provide for a further extension to fourteen days on due cause being shown.

Rule 22 describes the report. Incorrect reporting, as it opens a wide door to fraud, is made penal.

Rule 23 carries on the process to registering and marking the timber. These registers are the great safe-guard of the trade. Here can be seen, for years past, all the timber and its marks that any individual has brought down; and extracts of such registers are valued as affording a clue to all sorts of transactions.

Rule 24.—Registration being made, and the report having lain open without objection for the prescribed period, the "teingzá" or certificate is granted: it is a paper merely asserting that a certain person has, without objection, reported as his, and registered, and possesses actually lying at Kadoe or Nantay, so much timber. No attempt is made by these rules, or in practice, to give any fictitious value to these "teingzás." They are not evidence of ownership, nor are they admissible as evidence of anything unless the ordinary law of evidence makes them so; but they are regarded in practice as reliable proof that the person holding them is really in possession, and in all probability, seeing the great chance of detection were it otherwise, has a good title to the timber. The trade consequently accepts them as a sort of title-deed, and money advances, &c., are made on the simple deposit, with or without endorsement, of the papers as security,— the deposit constituting, in fact, a sort of equitable mortgage of the timber.

Rule 26 needs no remark.

Rule 27 provides for the grant of extracts from the register, which are useful to people who are anxious to have a clue to past timber transactions.

Rule 28 sees the timber finally paid for and passed out.

Rule 29 declares the Government lien on all timber for revenue, duty, and other dues. This subject has already been alluded to; it ought to be clearly laid down, to obviate any conflict of the authorities.

Rule 30 is designed to protect the officers of Government from dealing with agents whose authority is limited in such a manner as to involve doubt or complication. Sometimes it happens that an owner gives an agent authority to dispose of a limited number of logs at Kadoe; this would involve the controlling officer in intricate duties of account and calculation, and it is but right that he should have the power to refuse to enter into such a position.

<h3 style="text-align:center">CHAPTER V.</h3>

Of the management of the Kadoe Timber Station.

Rules 31 *to* 39.—This chapter may be dealt with as a whole.

It provides for the government of the station, which consists of a plot of land on either side of the river, and of islands between.

It is portioned out into allotments, on which timber is stored. Forest-guards watch the entrance and exit of all timber. The grant of the allotments is regulated, and provision is made for the maintenance of a watchman on each allotment, who is the paid servant of the timber owners, but is nevertheless subject to the orders of the Forest Officer in the manner defined. A rule (No. 37) guards against responsibility being laid on Government for the safety of the timber, but describes the duty of the Kadoe officials in using diligence to protect property from loss.

<h3 style="text-align:center">CHAPTER VI.</h3>

Of the Duties of Forest Officers under Act XIII of 1873.

Rules 40 *to* 48.—The fact that the management of rivers and timbers is vested in the officers of the Forest Department has been already intimated by Rule 2. In order to reduce the subject to a system, and enable officers to be checked, a scheme of duties for every grade is prescribed.

Rule 45 provides for cases not unfrequent, when a superior officer happens to be on the spot, and it is advisable that he should act at once, perhaps, in a sphere in which a subordinate would be the proper person to act were he present.

Rule 46 also makes preventive action the duty of all Forest Officers. A man belonging to beat A, who sees an offence committed in beat B, is not entitled to pass it over, saying, "I do not belong to beat A, it is no business of mine."

Rule 47 defines what officers are entitled to complain and prosecute before a Magistrate: this question has often been raised, and has hitherto been nowhere authoritatively laid down.

Rule 48 contains the usual prohibition against Forest Officers engaging in trade, &c.

CHAPTER VII.
Of Penalties.

The subject of penalties has been incidentally alluded to before. It is necessary to provide specially for the process of arresting (for the general provisions of the Criminal Procedure Code, though dealing with such powers under special laws in a general way, would not allow the arrest without warrant). Now it is *absolutely essential* that in dealing with these timber cases the power of immediate arrest should be granted. Once let it be known—and it *always is* known—that a warrant has been asked for, and all traces of the crime will disappear like magic.

Rule 50.—The general power of confiscation is conferred by the Act itself in Section 6; but it seems needful to develope the subject by describing and regulating in detail the formal procedure necessary to indicate the seizure which is preliminary to confiscation. The seizure must extend to implements and animals used in the offence (just as in the Forest Act), for in nine cases out of ten the *timber* itself, not being in any sort of sense the property of the person from whose hands it is taken, "confiscation," as far as *it* is concerned, is practically a misnomer.

If the seizure and confiscation of such articles is deemed *ultra vires* as regards the power conferred by Section 6, it is hoped that it may be deemed one of the "such other matters connected" as the Act permits the Chief Commissioner to regulate.

Rule 51 provides a necessary description (copied *verbatim* almost from the corresponding section in the Forest Act VII of 1865) of the hearing and conduct of applications before a Magistrate for confiscation, and for the consequent vesting of the confiscated property in the Crown.

Rule 52 continues the subject by providing an appeal when the property is over the limited value of Rs. 100, which is a question of fact easy to be determined.

Rule 53 provides a penalty for the breach of rules,—a subject already noticed. In a great majority of these cases, the offender is of that class where a mere money fine is of no deterrent effect; a short imprisonment, even of a day or a week, is the thing really feared; the penalty is therefore purposely prescribed as optionally embracing either mode of punishment or both.

The *proviso* is a mere declaration of what is obviously the existing law; but objections have been so often raised that it seems desirable to prevent, by a positive declaration, so idle a plea.

Rule 54 is possibly unnecessary in view of section 539 of the Criminal Procedure Code; but it completes the subject, and, being correct in itself, does not seem open to objection.

Rule 55 is a necessary addition. We are greatly dependent on information for the detection of those hardly-detected and easily-evaded offences, and a reasonable encouragement, without amounting to any inducement to fabricate complaints, is called for.

In conclusion, it only remains to be noted that it is presumed that the rules having received the sanction of the Governor General in Council, will have the "force of law."

The writer has no means of looking to see whether the "general clauses" or other Act render such a declaration in the present Act (XIII of 1873) necessary.

There is also no specific power of amendment of the rules given; but as the Act empowers the Chief Commissioner to make rules "from time to time," it seems that this includes the power of substituting a new or altered rule for any one of the present Code that time and practice—the only ultimate tests of legislation—may indicate as necessary.

The declaration of the rates of duty, &c., is left for a separate notification under section 3 of the Act.

RANGOON; }
The 5th September 1873. }

B. H. BADEN-POWELL,
Officiating Inspector General of Forests.

E.—DRAFT FOREST RULES FOR BRITISH BURMA, 1873.

THE following rules for the better management of the Government forests in British Burma, drawn up under Act VII of 1865, have been confirmed by the Governor General in Council, and are, in accordance with Section 6 of the Act, published for general information.

CHAPTER I.

Of the classification of Government Forests.

1. All Government forests which are maintained for the supply of timber and other forest
produce, or for the purposes of shade and protection, are
called "forest reserves."

Three classes of Government forest.

Forest reserves are either—
(1)—Ordinary reserves.
(2)—Special reserves.
(3)—District reserves.

"*Ordinary reserves*" are those which are maintained primarily for the preservation of
natural moisture, the retention of the soil on mountain
ranges, and for protection against landslips; for the pre-
servation of the banks of streams, rivers, and other waters, and for the maintenance of the
water-supply therein; for the protection of land against shifting sands, and damage by flood
and diluvion. They are in a secondary degree only intended for the supply of timber and
other forest produce; they are subject to such general protection as is hereinafter prescribed,
suitable to maintain them in a state sufficient for their purpose as above defined.

"Ordinary reserves."

"*Special reserves*" are those which are maintained primarily for the production and yield
of timber, wood, fuel, and other forest produce for the
benefit of the State and the supply of the market, and
further to fulfil the purposes of the *ordinary reserves*. They are subject to such special protection, and to provisions for their reproduction, treatment, and working, as are hereinafter
prescribed, suitable to maintain them in the condition of the greatest and most beneficial permanent yield.

"Special reserves."

"*District reserves*" are those which are maintained
for the purposes following :—

"District reserves."

(a)—Forests and plantations in the vicinity of towns and villages, and which, though they
are the property of Government, are designed to supply timber, wood, fuel, and
other forest produce for the use and convenience of the inhabitants of such towns
and villages.

(b)—Groves, plantations, and avenues for ornament or shade in the vicinity of roads,
public buildings, and places of public resort, or for the maintenance of the
banks of springs, streams, tanks, and other waters, and of the water-supply
therein.

(c)—Plantations in course of formation by district authorities, designed, when grown up,
either to form *district reserves* of the classes (a) or (b), or to be treated as
special reserves.

2. The administration and control of *ordinary* and
special reserves is vested in the officers of the Forest
Department.

Administration of the forests.

3. The administration and control of *district reserves* is vested in the civil and revenue
authorities of districts, subject to such general or
special provisions regarding inspection and advice
as are contained in these rules, or as the Local Government may from time to time prescribe.

Administration of district reserves.

4. A list of the names of forests constituting the *ordinary* and *special reserves* for the
time being shall from time to time be published in the
official Gazette. And lists of all *ordinary*, *special*, and
district reserves in each civil district shall be maintained, and be open to public inspection in
each civil district office and in each forest divisional office.

Forest lists to be published.

5. In these rules "minor forest produce" means and includes bark, leaves, grass, bamboos,
wax, oil, resin, gum, varnish, honey, tusks, horns, and
skins; seeds, fruits, stones, lime, and every kind of forest produce except wood and timber.

"Minor forest produce."

CHAPTER II.

Of Ordinary Reserves.

6. Whenever it appears to the Conservator of Forests desirable that a tract of Government
forest should be maintained and treated as an *ordinary*
reserve, he shall consult with the civil district officer
or officers within whose district the reserve, or part of it,
is situate, and shall submit to the Local Government a report prepared jointly by him and such
officer, setting forth the local name, position, estimated extent (if not otherwise known), and
natural boundaries of the forest, and the reasons for its treatment as such reserve; and if the
local Government is satisfied—

Procedure in demarcating an ordinary re-serve.

(1)—that the tract is suitable for the purposes of an *ordinary reserve* as defined in
rule 1, and

(2)—that its reservation will not offer undue impediment to the spread of population and
cultivation of waste land, and will not cause hardship to existing villages and
communities as regards the supply of forest produce for purposes, domestic and
agricultural,

it may sanction the formation of such reserve.

7. Pending the settlement of question of reservation, a preliminary notice shall be issued through the officer of the district in which the proposed reserve, or any part of it, is situate, prohibiting any new toungya clearings, any new cultivation, establishment of new houses or villages within the proposed boundaries, until the orders of Government are issued.

Preliminary notice to maintain the status quo.

8. On sanction being given, the Conservator of Forests shall determine whether demarcation by pillars or other artificial marks is necessary, and if it is, shall proceed to cause the demarcation. A final proclamation shall be then issued and made known in the vicinity through the officers of the district or districts in which the reserve, or any part of it, is situated, setting forth the boundaries of the reserve, and warning all persons against the acts prohibited by Rule 9.

Demarcation and final notice.

9. The following acts are prohibited in *ordinary reserves :—*

I.—Girdling, felling, lopping, stripping off leaves, barking, burning or otherwise injuring and cutting up or dragging trees or timber of the valuable kinds following, that is to say—

Acts prohibited in ordinary reserves.

Kyoon (teak)	...	(*Tectona grandis.*)
Thitkâ or Kathitkâ	...	(*Pentace Burmanica.*)
Thitkado	...	(*Cedrela Toona* and varieties.)
Thingán and its varieties	...	(*Hopea odorata, &c.*)
Yemmanny	...	(*Gmelina arborea.*)
Thitsê	...	(*Melanorhæa usitatissima.*)
Kanyen	...	(*Dipterocarpus alata.*)
Thinwin	...	(*Rosewood.*)
Shaw	...	(*Sterculia sp.*)
Pyinma and Laihza	...	(*Lagerstræmia.*)
Kokoh	...	(*Albizzia Lebbek.*)
Pingado	...	(*Xylia dolabriformis.*)
Padouk	...	(*Pterocarpus dalbergioides.*)
Shah	...	(*Acacia Catechu.*)
Eng	...	(*Dipterocarpus tuberculata.*)
Hmâyâ	...	(*Grewia microstemma.*)
Bambway	...	(*Careya arborea.*)
Yindeik	...	(*Dalbergia cultrata.*)
Anân	...	(*Fagræa fragrans.*)

without permission according to the rules contained in Chapter V. The local Government may from time to time add other species to the list.

II.—Setting fire to the forest, or negligently allowing fire to spread to the forest; negligently lighting or leaving alight camp fires; dropping embers on the ground; kindling kilns or burning lime or charcoal without permission under Rule 39. Hunters are prohibited from setting fire to the forest.

III.—Clearing toungyas, except with permission, as in Rule 17, Clause VI.

IV.—Injury to the soil by removing turf, vegetable mould (humus), dead leaves, and the like.

V.—Damage, negligent or wilful, by elephants, buffaloes, bullocks or otherwise, in process of working or extracting forest produce.

VI.—Sawing up or conversion of timber, burning, marking or defacing, or altering marks on timber or trees in the forest without permission of the officer in charge of the forest division.

VII.—The collection of minor forest produce, *other than* grass and bamboos, shall be effected only with permission previously granted as hereinafter provided.

10. The Conservator of Forests may (when such a course is necessary) close any part (giving due notice thereof) of any ordinary reserve against grazing, pasturing, and trespass by cattle.

Closing forest against grazing (ordinary reserve).

11. Except as above provided, the use and collection of wood and forest produce is free and unrestricted, provided also that the produce is taken for the use of persons collecting it, and not for export or merchandise, and provided that it is taken in such reasonable quantity as not to denude any part of the forest, or endanger the object of the reserve.

12. The proper treatment of "*ordinary reserves*" shall be provided for by such general working plans as may be necessary; such plans in all cases fixing the quantity of timber removable otherwise than as single trees or for petty demands, if there is a stock suitable for such removal. *Ordinary reserves* shall be fire-traced where possible, and otherwise protected from the ingress of fire, as the Conservator of Forests may direct.

Working plan for ordinary reserves.

CHAPTER III.

Of Special Reserves.

13. Whenever it appears to the Conservator of Forests desirable that any Government forest should be treated and maintained as a *special reserve*, he shall submit a proposal to the local Government stating the position, estimated extent (if not otherwise known), local name and boundaries

Procedure in demarcating a special reserve.

of the proposed *special reserve*, and its forest capabilities; and if the terms of such proposal satisfy the local Government—

(1)—that the produce of the tract is sufficiently valuable, *or* that its soil and other conditions are such as to render it capable of supporting valuable produce, and that its extent is such as to make it worth while to preserve;

(2)—that its position as regards export lines, vicinity of markets, and the like, is favourable;

(3)—that no rights exist within it, *or* if they exist, a suitable arrangement for their extinction by exchange or compensation will be agreed to by the right holders;

or that the rights can conveniently be expropriated by the law for the time being in force;

or that the rights have been ascertained correctly and defined in writing, and found to be such as do not interfere materially with the proper treatment of the forests,

then the formation of such special reserve may be sanctioned. *Provided* that if the proposed special reserve is not included in an area already sanctioned as an ordinary reserve, the provisions of Rule 6, regarding consultation with the district officers and the submission of a joint report, shall be complied with.

14. Pending the settlement of the question of reservation, a preliminary notice shall be issued as provided in Rule 7.

Preliminary notice to maintain status quo.

15. On sanction being given, the Conservator of Forests shall, if the reserve is not sufficiently demarcated by permanent roads, streams, or other existing boundaries or landmarks, cause the same to be marked out by distinct and permanent boundary marks bearing a serial number.

Demarcation.

A final proclamation shall then be issued through the officers of the district or districts in which the special reserve, or any part of it, is situate. Such proclamation shall be publicly made known in the vicinity of the reserve, and shall indicate the boundaries of the reserve, and shall further warn all persons against any trespass or infringement of the forest rules regarding such reserves.

Final proclamation.

16. The Conservator of Forests may, with the sanction of the Local Government, determine what roads and pathways through a special reserve shall be authorized for public traffic, and may close all other tracks and pathways, giving notice in the vicinity of such closing.

Power to close useless tracks and pathways.

17. The following acts are prohibited in special reserves :*—

Acts prohibited in special reserves.

I.—Setting fire to the forest, or kindling any fire in the vicinity thereof in such manner as to endanger its safety.

Lighting camp fires, except on cleared camping places, which will be prepared for the purpose, where the extent of the reserve renders it necessary for travellers to halt in it.

Burning lime, charcoal, or bricks in the reserve, or so near to it as to endanger the forest, without permission under Rule 39.

II.—Kindling or carrying any fire between 1st February and 1st June, except at cleared halting places as aforesaid.

III.—Trespass by men, elephants, or cattle, off the authorized roads and pathways.

IV.—Girdling, felling, lopping, burning, injuring, stripping off leaves, barking, or tapping for oil, varnish, or resin, any tree, and moving or dragging any timber otherwise than at a regularly authorized working by virtue of a lawful order or permit according to Rules 32 to 35, as the case may be.

V.—Injury to the soil by removing turf, vegetable mould (humus), and the like.

VI.—*Toungya* clearing and every form of cultivation, except by written order of the Conservator of Forests, which may be issued to allow the same—

(a).—With a view to subsequent cultivation of teak or other valuable forest produce.

(b).—For the support of labourers located in the forest.

(c).—For the support of subordinate forest officials located in the forest.

Breach of any conditions entered in such order, for the safety of the forest, shall be deemed to be a breach of this rule.

VII.—Sawing up or conversion of timber, or burning, marking, defacing, or altering marks on timber or trees in the forest without the permission of the officer in charge of the forest division.

VIII.—Grazing or pasturing of cattle, elephants or pigs, except by permission of the officer in charge of the forest division under Rule 40.

IX.—Damage, negligent or wilful, by elephants, buffaloes, bullocks or otherwise, in process of working or extracting forest produce.

X.—Minor forest produce shall be collected only by lawful permit issued under Rules 35 to 38, as the case may be.

18. Special reserves shall in all cases be efficiently fire-traced with such external and internal cleared lines as the Conservator of Forests may deem necessary, and shall be fenced in places where, from the vicinity of a frequented road or otherwise, trespass by men or cattle is especially apprehended.

Fire-tracing of special reserves compulsory.

* Special reserves being a small area under planting and other 'intense' treatment, it is absolutely necessary to prohibit *all* sorts of trespass; above all, to keep out men and cattle from trampling down the young seedlings.

19. The local Government shall prescribe for the special reserves, either singly or in such *Detailed working plans.* groups as may be deemed proper, or in connection with ordinary reserves, a working plan which shall define—

(1.)—The system of general treatment and reproduction, whether for high forest, coppice, or otherwise.

(2.)—The method of exploitation and utilization of material, including the regulation of pasture, and of the collection of minor forest produce.

(3.)—The system of artificial cultivation of trees, where necessary to resort to it.

(4.)—The systematic distribution of the various works to be carried out over the appropriate years of the different periods.

The working plan shall make provision for its periodical revision.

20. In any case in which the exploitation of timber is effected by permit-holders having *Working plan, provisions to be given to permit-holders.* license to cut and remove timber under Rule 35, the permit-holder shall be furnished with such certified extracts from the working plan as may be necessary, and the same shall be terms of his permit or agreement.

CHAPTER IV.

Of District Reserves.

21. Every officer of a district shall so arrange the issue of grants to cultivate waste land *Reservation of lands for district forest.* (under the law or rules for the time being in force) as to secure to each town, village or group of villages, such area of waste land (to be kept as forest) as shall be convenient and suitable to the objects of a district reserve of the class (a) described in Rule 1.

22. The district officer, acting under such general or special instructions as the Commis- *Supervision.* sioner of the division may issue, shall make such arrangement for the supervision and protection of these reserves as may be necessary to fulfil their object, and may entrust myookes, thoogyees, village headmen, and other subordinate officials with the duty of protecting and supervising them.

23. Valuable trees of the kinds specified in Rule 9, except cutch (*Acacia Catechu*), shall *Valuable trees not to be cut.* not be cut in any way, nor injured, nor the timber thereof dragged or removed, and boat hulls shall not be hollowed out without the orders of the Deputy Commissioner, or some one authorized by him to grant such orders, and on payment of such fees as may be prescribed in each district by order of the Local Government.

24. Cutch trees (*Acacia Catechu*) not being included in any special or ordinary reserve, *Cutch how dealt with.* shall be dealt with according to the provisions of the revenue law with reference thereto.

25. The kanyen and thitsi trees in each district shall be grouped into convenient circles, *Oil and varnish trees how protected.* and the right of preserving and working them for wood-oil and varnish shall be sold by triennial lease by the Deputy Commissioner.

The right of collecting honey and bees' wax shall be leased or granted according to the practice heretofore prevailing in each district.

26. Except as provided in the preceding rules, all forest produce in district reserves shall *Forest produce is free otherwise.* be free, provided it is used locally for the domestic or agricultural wants of the persons taking it, and not for purposes of mercantile profit.

27. No trees shall be felled, lopped, burned, tapped, cut or injured within fifty feet of *Protection of tree and roadsides.* either side of any public road, or of any stream, canal or creek. The Deputy Commissioners of districts (subject to such general or special instructions as the Commissioners of their divisions may see fit to issue) shall determine and notify publicly what roads, streams, creeks, and canals in their districts are subject to this rule.

Subject to similar instructions, the Deputy Commissioners of districts shall determine *And at public places.* what groves of trees shall be maintained in the vicinity of springs, wells, pagodas, zayats, and public buildings, and no tree in groves so maintained shall be felled, lopped, or injured without the express orders of the Deputy Commissioner.

The provisions of this rule may be also applied by order of the Local Government to railways, and river or other embankments.

28. The Deputy Commissioners of districts may cultivate suitable tracts of Government *District plantations.* waste land as plantations of valuable trees. When formed and completed, the Local Government shall determine whether such plantations shall be kept as district reserves or managed as special reserves.

29. Such plantations shall ordinarily be made under charge of the thoogyees, and a commission at such rate and on such area or number of trees successfully reared as the Local Government shall prescribe, shall be payable by the Deputy Commissioner to the thoogyee in charge on certificate of the Forest Officer appointed in that behalf that he has inspected, and is satisfied with the work.*

---

* It is contemplated that the fees, &c., derived from district forests should be formed into a 'district planting fund,' out of which such commission and other charges could be paid; but I do not think the constitution of a fund is one of the subjects allowed to be brought into the rules. So it must be arranged executively.

Chapter V.

Of the working out of Timber and other Forest Produce in Special and Ordinary Reserves.

A.—Of the extraction of timber and trees.

30. The rules in this chapter apply to ordinary and special reserves only, unless the Local Government shall otherwise direct.

Method of obtaining timber.

31. Application to fell and remove single trees, or any number not exceeding ten, of the "valuable kinds" mentioned in Rule 9 in the case of ordinary reserves, or of any kind whatever in the case of special reserves, shall be made to the officer in charge of the forest division, who may issue a *"special permit"* on payment of such fee, for each tree, as may be prescribed by the Local Government.

Special permit for trees.

The officer in charge of the division may delegate authority to grant special permits to any officer in charge of a forest sub-division, and may delegate authority to grant special permits for single trees only, to any Forest Ranger in charge of a forest range.

32. All such special permits shall be issued uniformly from a book furnished with a counterfoil, and shall specify the number of trees, the kind, the fee paid, the place of cutting, and the time within which the trees must be felled and cleared out.

Form and conditions of special permits.

The pre-payment of fees is in all cases compulsory. The special permit lapses absolutely on expiry of the time fixed, whether the trees have been felled and removal completed or not; and the trees, if not cut, and the timber thereof, if felled but not removed, revert to Government, but no person shall be entitled to recover any fee paid for the permit.

33. An officer empowered to issue special permits may in all cases refuse the application whenever the working plan does not contemplate trees of the kind applied for being felled, or whenever he considers that the condition of the forest and the quantity of trees available warrant such refusal, or he may grant such smaller number than that applied for as he thinks fit.

Restrictions on the issue.

34. In case of requisitions for more than ten trees, or in the case of a regular exploitation of teak or other timber under the working plan, the felling and removal must be made according to the system in force in the forests at the time, and according to the working plan.

Working out trees in large numbers, &c., how effected.

Such system may be either—

(1).—Felling and removing timber by permit agreements in such form and in such terms as the Local Government may direct.

(2).—Felling timber by Government agency direct, and disposal of such timber either in the forest or at timber depôts after extraction and floating down.

B.—Of the utilization of minor forest produce.

35. The right to collect or utilize any kind of minor forest produce (the collection of which without permission is prohibited by Rules 9 and 17) may be granted either by "special permits" for limited quantities, or applying to certain portions of the forest, and holding good for short terms only, or by "general leases" or permit agreements, for a term of years.

Collection of minor produce.

36. Special permits may be issued by the same authority, in the same form, with similar specifications as to time, locality, sort of produce, and fee paid, and under the same conditions as special permits for trees under Rule 32, and they may be refused or restricted as provided in Rule 33.

Special permits.

The amount of fees shall be according to a scale promulgated from time to time by the Local Government.

37. General leases or permits to collect minor forest produce for a term of years may be issued by the Conservator of Forests on such conditions as he thinks fit. Such leases or permits may be either sold by auction or tendered for, as the Conservator of Forests may from time to time direct.

General permits.

38. Free permits, either special or general, may be issued to the headmen of Karen and other villages, in the neighbourhood of any ordinary or special reserve, authorizing the villagers to collect such produce for their own local use.

Free permits to certain villagers.

Such permits may be withdrawn by the Conservator of Forests in case of breach of forest law, or of the conditions entered in the permit.*

39. The burning of lime, charcoal, or bricks in ordinary or special reserves shall only be allowed by virtue of a written order from the officer in charge of the forest division. Such order shall contain the necessary conditions regarding fire-tracing and protection of the forest, and regarding the sort and quantity of fuel to be used for the burning, and how it is to be obtained.

Burning kilns.

* These permits might be conditioned to give the Kyeydangyen, &c., the responsibility of taking care of the forest and of preventing fire and wanton damage as far as possible. We give them the free run of the reserved forest as well as the whole unreserved tract which is open to them, and it is not too much to ask in return that they should help to prevent damage.

Fees for each kiln may be charged as the Conservator of Forests may direct.

Breach of any condition of the order shall be deemed to be a breach of the prohibition in Rules 9 and 17, as the case may be.

40. Permits to graze cattle may be issued in special reserves, where consistent with *Grazing permits.* the working plan, by the officer in charge of the forest division only.

Such permits shall be in writing, and shall specify the number and sort of cattle, the part of the forest to which they apply, the fees payable, if any, and the name of the person or persons who are to be held responsible in case of any damage or trespass contrary to the terms of the permit.

41. All permits or orders other than those described in Rule 84 or Rule 37 shall be *Return of expired permits.* returned to a Forest Officer when the permit has been satisfied, or when the time of it has expired.

Permits not transferable. 42. No permit, or lease, or other written order, issued under this chapter, is transferable, except under special conditions entered therein.

CHAPTER VI.

Of Forest Officers and their Duties.

Forest establishment. 43. The forest establishment is divided into "con-
Official titles. trolling" and "subordinate." The controlling establishment consists of—

The Conservator	...	...	... (Thit-tau-woongyee).
Deputy Conservators	...	...	... (Thit-tau-woon).
Assistant or Sub-Assistant Conservators		...	(Thit-tau-woondouk).

The subordinate establishment consists of—

Forest Rangers	...	...	... (Taw-oke).
Foresters	...	...	... (Goungwai).
Forest Guards	...	...	... (Taw-soung).

When such subordinates are posted to duty at timber stations or depôts, they may receive such addition to their official designation as may serve to indicate such duty.

No others to be used. No other titles or designations shall be used or recognized officially.

Forest "division." 44. A "forest division" is the charge ordinarily held by a Deputy Conservator or senior Assistant Conservator.

"Sub-division." A "forest sub-division" is a charge subordinate to a division, ordinarily held by an Assistant Conservator.

"Range." A "forest range" is a charge subordinate to a subdivision, or directly to a division, as the case may be.

"Beat." A "forest beat" is a charge subordinate to a range, and is held by a Forester, assisted by one or more forest guards.

No other charges recognised. No other local division of charges shall be officially used or recognized.

45. The appointment, posting, degradation, and dismissal of the controlling establish-
Appointment and control. ment rests with Government, and of the subordinate establishment (within the scale and rates of pay sanctioned by Government) with the Conservator of Forests, or with such officers in charge of forest divisions as regards subordinates in such divisions, as may be authorized in that behalf by the Conservator of Forests.

46. The Conservator of Forests may suspend any Forest Officer (subject to report to the
Power of suspension. Local Government in the case of the controlling establishment) pending inquiry into any official misdemeanour or criminal charge of which such officer is accused. The officer in charge of a forest division may exercise similar powers in respect of the subordinates in his own division (subject to report to the Conservator of Forests).

47. The official uniform and badges to be worn, and weapons to be carried by the forest
Official uniform and weapons. establishment, shall be determined by the Local Government. Every member of the subordinate establishment shall wear some distinctive mark of his functions and authority when on duty.

48. The "Conservator of Forests" has the administrative charge and supervision of the
Duty of Conservator of Forests. whole of the forests and works managed by the Forest Department, and of the establishments provided for them. He is charged with the control of the record of his office and other property of Government thereto pertaining. He is also charged with such duties of correspondence and examination and control of the provincial forest accounts as may be prescribed by the Local Government, or the Government of India. He shall further inspect and report, or advise on all such matters, as the Local Government may require.

Duty of a Deputy Conservator. 49. A "Deputy Conservator" has executive charge of a forest division, as defined in Rule 44, or of some special branch of work.

When in charge of a division, he shall be deemed to be *entrusted* with the Government property in the timber and other growing stock of the forests in the same manner as he is with the cash, stock, and other property in his charge.

It is also his duty—

(1).—To keep all accounts, and to make all reports and returns, and to take due care of property in his charge; also to keep up such diaries and note-books, and conduct correspondence in such manner as may from time to time be prescribed to him.

(2).—To use his utmost endeavours that the forest rules are not infringed; that the working plans and general orders regarding the preservation, treatment, exploitation, and reproduction of the forests in his division are efficiently carried out; and to see that all subordinates are at their posts and do their duty. He shall also report all serious cases of forest offences, or fire or damage to the forests, to the Conservator of Forests, and shall take the necessary steps to bring offenders to justice, and repair the damage done.

(3).—To take proper steps to realize all royalty and other dues of Government.

(4).—To make such special inspections and reports on any matter connected with forest work as the Conservator of Forests may require.

Duty of an Assistant or Sub-Assistant Conservator. 50. An "Assistant Conservator" in charge of a division has the same duties as above prescribed for a Deputy Conservator.

An Assistant or Sub-Assistant Conservator in charge of a sub-division is entrusted with the Government property in the growing stock of the forest in the same way, and has the same duties in the sub-division and the charges attached thereto, as a Deputy Conservator in a division, but subject to the lawful control and direction of the Deputy Conservator, and in correspondence with him. If not in charge of a sub-division, but specially or generally employed otherwise, he shall carry out his duties and employment under the orders of the Deputy Conservator of the division in which he is employed, or directly under the orders of the Conservator of Forests, as the said Conservator may direct.

Duty of a Forest Ranger. 51. A Forest Ranger, if in charge of a range, shall be deemed to be entrusted with the Government property in the timber and other growing stock of the forests in his range.

It is also his duty—

(1).—To live in or close to the forest, to be acquainted with every part of his range, and keep himself informed of what goes on in it; to report to his immediate superior the occurrence of any fire, forest offence, or damage to the forest.

(2).—To visit and report to his immediate superior the state of all the boundaries and boundary marks in his range at least twice a year.

(3).—To prevent by every lawful means in his power the commission or continuance of any breach of the law or forest rules, and to arrest the offender according to Section 8, Act VII of 1865.

(4).—To prevent forest fires by every lawful means in his power, and in the event of one occurring to procure forthwith the needful help, and use every effort to extinguish it. Before and during the usual season of fires, he must carefully see the fire-paths and tracings are clear, and kept clear of all inflammable matter.

(5).—To seize all timber, wood, and other forest produce unlawfully cut, moved, collected, or reasonably suspected to be so, which he shall find in the forest or in transit therefrom, and deal with it, subject to such orders as he may receive from his divisional officer, according to Sections 8 and 11 of Act VII of 1865.

(6).—To see that all works of cutting or exploitation of timber, or collection of forest produce, grazing, cultivation, and all other works carried on in the forest, by virtue of the working plan or lawful orders of competent authority, are duly and properly executed without injury to the rest of the forest.

(7).—To carry out all lawful orders of the officer to whom he is subordinate; to send in such returns and reports, and such accounts of all money entrusted to him as he may be required to do.

(8).—To maintain a diary showing his journeys, occupation, and all noteworthy incidents, and to submit the same to his superior officer at such intervals as may be prescribed. When on special duty or employment, he shall execute such duty according to the lawful orders he may receive.

Duty of a "Forester." 52. A *Forester*, if in charge of a *beat*, has the following duties:—

(1).—To live in or close to his beat; to patrol it all over frequently so as to know it thoroughly, and know everything that goes on in it.

(2).—To visit and report on the boundary marks in his beat at such intervals as he may be directed.

He has also the duties described in paragraphs 3 to 7, both inclusive, of a Forest Ranger's duty as far as regards his own beat.

A Forester in charge of any special work, or employed in any special work under a superior officer, shall carry out the work according to the orders given, and act according to the lawful directions of his immediate superior, as the case may be.

53. A *forest guard* has ordinarily to accompany and assist a Forester in his beat.

Duty of a "forest guard." He must live in or near the beat, and be ready at all times to assist the Forester in his duties, and accompany him when patrolling. When the Forester is not with him, he has the same duties as a Forest Ranger in respect of forest offences and fires (paragraphs 3 and 4), seizure of unlawful produce (paragraphs 4 and 5), and supervision of works (paragraph 6).

It is his duty at all times to obey the lawful orders of his superior officers. When employed on any special work, he is to assist in such way as he may be lawfully required.

54. All subordinate officers in charge of special reserves shall regard it as a duty to be **Duty of all subordinate officers regarding forest improvement.** constantly on the watch, while out patrolling, to do everything in their power for the well-being of the forest.

They must *always* be armed with a suitable knife to cut creepers and parasites, prune off broken boughs, cut down teak saplings which have been trampled or broken by accident, to destroy noxious animals and insects; and during the proper season they should carry seed of teak and other valuable trees so as to be able to dibble it into the soil in blank places.

Controlling officers will specially notice the indications of this rule being attended to when visiting the forest, especially at *visits not previously announced.*

55. All Forest Officers, wherever they are, and on whatever duty or employment, have **Forest Officer's duty to prevent offences wherever he is.** always the duty of preventing any offence against the forest law or rules, arresting the offender according to Section 8, Act VII of 1865, extinguishing forest fires, and protecting the property of the State or of private persons while detained at or in charge of any Government timber depôt or station.

But in such cases they shall communicate what they have done to the officers in local charge as soon as may be, and the local officer shall take such further steps as the matter in hand may require.

56. Any Forest Officer may lay a complaint to the Magistrate or to the police, as the **Power to complain and prosecute.** case may be, apply for a confiscation under Section 11, Act VII of 1865, in any case of the breach of the law occurring in his own local charge. But the officer in charge of the division may direct any other officer subordinate to him to complain or prosecute, or may conduct the case himself.

57. Any Forest Officer of a superior grade may in case of necessity exercise all or any of the functions of the grades below.

58. No Forest Officer shall trade or engage in any agency business, or in any work or **Prohibition against trading.** employment other than his duties as such Forest Officer, except with the permission in writing of the Conservator of Forests.

CHAPTER VII.

Penalties.

59. For every act or omission contrary to any of these rules, or abetment thereof, and **Penalty.** for every wilful neglect of duty on the part of any Forest Officer of the subordinate establishment, a fine not exceeding Rs. 500, or, in default of payment, imprisonment as provided in Section 67 of the Indian Penal Code.

But nothing in these rules is intended to prevent the prosecution of any person for any offence under the Indian Penal Code or other law to which he may be liable, and not under these rules.

60. One-half the fine awarded on any conviction for a forest offence may be awarded by **Reward to informer.** order of the Magistrate trying the case to the informer or other person in consequence of whose aid or information the offender was brought to justice.

IX.—AJMERE FOREST REGULAION.

NOTIFICATION—By the Government of India, in the Foreign Department, No. 304R., dated 23rd December 1874.

A REGULATION TO PROVIDE FOR THE ESTABLISHMENT OF STATE FORESTS IN AJMERE AND MHAIRWARRA AND TO PREVENT THE INDISCRIMINATE FELLING OF TREES AND REMOVAL OF JUNGLE IN MHAIRWARRA.

Preamble. WHEREAS by a resolution passed by the Secretary of State in Council on the sixteenth day of March 1871, the provisions of the thirty-third of Victoria, Chapter three, Section one, were declared applicable to Ajmere and Mhairwarra:

And whereas the Chief Commissioner of Ajmere has proposed to the Governor General in Council a draft of the following Regulation, together with the reasons for proposing the same:

And whereas the Governor General in Council has taken such draft and reasons into consideration, and has approved of such draft, and the same has received the Governor General's assent:

In pursuance of the direction contained in the said section, the said Regulation is now published in the *Gazette of India,* and will be published in the local Gazette, and will thereupon have the force of law :—

Preliminary.

Short title.
1. This Regulation may be called "The Ajmere Forest Regulation, 1874."

Interpretation-clause.
2. In this Regulation, unless there be something repugnant in the subject or context—

the expression "villagers" includes the members of the proprietary body of any village; and any other persons or class of persons who may, by a written order of the Commissioner, subject to the control of the Chief Commissioner, be declared entitled to the status of villagers under this Regulation;

the expression "Forest Officer" means any person or persons whom the Chief Commissioner of Ajmere from time to time appoints to exercise the powers and perform the duties hereby conferred and imposed on a Forest Officer;

and the expression "cattle" includes also elephants, camels, buffaloes, horses, mares, geldings, ponies, colts, fillies, mules, asses, pigs, rams, ewes, sheep, lambs, goats, and kids.

Taking up of land under this Regulation.

Declaration for taking up land.
3. Whenever it appears to the Chief Commissioner of Ajmere expedient that any tract of waste or hilly land comprised in the area of any village should be taken up by the Government for the purposes of a State forest, a declaration in the form given in Schedule A hereto annexed, or to the like effect, and describing the land by its boundaries, or otherwise with convenient certainty, shall be published in the local Gazette, and a copy of such declaration in Hindi, together with a written explanation in Hindi of the terms as hereinafter laid down on which the land is taken by the Government, shall be delivered to the lumberdars of the village.

Legal effect on such declaration.
4. Such declaration shall be conclusive evidence as to the nature and condition of the land and as to the expediency of taking it up;

and on its publication in the local Gazette the following consequences shall ensue :—

(a.) The proprietary right to the land shall vest in the Crown, and in lieu of all rights which any person may now have in or to such land, the rights hereinafter in that behalf mentioned shall be reserved to the villagers;

(b.) The Forest Officer may enter and take possession:

Rights created in favour of villagers.
(c.) Subject to the rules and limitations in the next following section provided, the undermentioned rights over the land may be exercised by the villagers, that is to say:

To cut grass.
(1.) A right to enter upon the land to cut grass thereon;

To cut wood.
(2.) A right to enter upon the land to cut such wood as is reasonably necessary for their household requirements and agricultural implements;

To use ways.
(3.) A right to use all such ways of a defined and permanent character over the land as were in use by them at the time the declaration was published and are still adapted for use.

Villagers' rights to be exercised under control of Forest Officer.
5. The rights vested in the villagers, under section four, shall be exercised subject to the control of the Forest Officer, who may, from time to time, among other things, and subject to an appeal to the Commissioner of Ajmere—

(a.) issue written orders, determining the seasons at which grass may be cut, and the mode of cutting it, and prohibiting the cutting of it in any part of the land where such cutting would tend to damage the trees there growing;

(b.) issue written orders determining the season when, and the place where, wood is to be cut;

(c.) stop any way across the land, and assign another way instead of it, provided that the new way set out by him be a reasonably convenient substitute for the way so stopped.

6. There shall be distributed among and paid to those who, previous to the taking up of the land, were interested therein, the following proportions of the net profits (if any) from time to time resulting from the State forest operations on the land, after deducting all expenses of management, namely, of profits from operations other than the working of mines and quarries—two-thirds; of profits from the working of mines and quarries—one-half.

The amount of such profits, the times at which they are payable, the persons entitled to participate in them, the shares claimable by such persons, and the mode of distribution, shall be determined by the said Commissioner, subject to the control of the said Chief Commissioner, by a declaration in writing, and such declaration shall be final and conclusive as against all persons concerned.

7. If the members of any village community, or any other persons entitled to a share of profits under such declaration, have interfered with or obstructed the State forest operations, or have not rendered such assistance to the Forest Officer as may be lawfully required of them, the said Chief Commissioner may direct that there shall be withheld from them a sum not exceeding one-half of the profits which would otherwise have accrued to them or to the village community of which they are members, and such sum shall be withheld accordingly, and shall be credited to the Forest Department.

Forfeiture of part of profits for misconduct.

8. When any land has been taken up for a State forest under this Regulation, no fine shall be levied in respect of any trespass by cattle thereon until the Forest Officer has efficiently protected that portion, in which grazing is prohibited, by fencing, or, with the Commissioner's sanction, demarcated it by conspicuous marks, which have been duly notified in the vicinity. But this section shall not apply where cattle have been wilfully caused to trespass by the owner or any person in charge of them.

No fine to be levied for cattle trespass on unprotected forest.

9. The Chief Commissioner of Ajmere may, by a notification in the local Gazette, make rules consistent with this Regulation for the management and protection of State forests created under the provisions herein contained, and may, by a similar notification from time to time, alter, add to, or rescind such rules. He may, in making any such rule, attach to the breach of it, in addition to any other consequences that would ensue from such breach, a punishment, on conviction before a Magistrate, of a fine not exceeding, for the first offence, fifty rupees, and for the second or any subsequent offence, one hundred rupees.

Power to make rules.

Relinquishment of land taken up under this Regulation.

10. Whenever it appears to the said Chief Commissioner that a tract of land taken up under this Regulation is no longer required for the purposes of a State forest, a declaration in the form given in Schedule B hereto annexed, or to the like effect, and describing the land by its boundaries, or otherwise with convenient certainty, shall be published in the local Gazette, and a copy of the same in Hindi shall be delivered to the lumberdars of each village within the area of which any portion of such land was originally included.

Declaration for relinquishing land.

11. After publishing such declaration, the Commissioner of Ajmere shall, as soon as conveniently may be, proceed to restore the land so disforested to the communities or persons to whom it belonged before it was afforested, so far as the change of circumstances will permit, and subject to such charges for works of permanent improvement effected by the Government as to the said Commissioner seems proper.

Restoration of land disforested.

For this purpose he shall issue an order in writing, specifying the communities or persons to whom each portion of the disforested land is to be restored, and their interests therein, and the nature and incidence of the charges thereon. Such order shall be binding and conclusive on all parties concerned.

Restriction of the right of felling trees and making charcoal.

12. The Chief Commissioner of Ajmere may, by a notification in the local Gazette, make rules for the prevention of charcoal burning and destruction of trees in the vicinity of the State forests or in other places where these practices may in his opinion be injurious. In issuing such rules due regard will be had to proprietary rights.

Chief Commissioner may make rules.

Recovery of fines.

13. The provisions of sections sixty-three to seventy, both inclusive, of the Indian Penal Code, and of section three hundred and seven of the Code of Criminal Procedure, shall apply to all fines imposed under this Regulation, or under the rules made in the exercise of the power given by section nine of the same.

Recovery of fines.

SCHEDULE A.

(See Section 3.)

Form of declaration for taking up land.

The waste (☞) hilly land below described being required for the purposes of a State forest is hereby under the orders of the Chief Commissioner taken up for such purpose, and the present declaration is made and published under the Ajmere Forest Regulation, 1874, section three.

SCHEDULE B.

(See Section 10.)

Form of declaration for relinquishing land.

The land below described being no longer required for the purposes of a State forest, is hereby under the orders of the Chief Commissioner relinquished, and the present declaration is made and published under the Ajmere Forest Regulation, 1874, section ten.

X.—BERAR.

NOTIFICATION—By the Government of India, in the Department of Revenue, Agriculture and Commerce, No. 520, dated 25th October 1871.

THE subjoined rules have been sanctioned by His Excellency the Governor General in Council, for the administration of forest lands in the Hyderabad Assigned Districts, and are published for general information :—

FOREST RULES FOR THE HYDERABAD ASSIGNED DISTRICTS.

PART I.

Preliminary.

Interpretation-clause. 1. These rules shall come into operation from this date.

2. In these rules, unless there be something repugnant in the subject or context—

Forest shall mean any tract of country covered with trees, brush-wood or grass, and shall include such waste or cultivated lands, roads, streams, and other waters, as may be situated within such tract.

The word *cattle* shall, besides horned cattle, include elephants, camels, horses, asses, mules, sheep, goats, and swine.

Forest rights shall include the right of ownership in trees, timber, bamboos, brush-wood, grass, or other produce of a forest; the right to prohibit the cutting, burning or using of trees, timber, bamboos, brush-wood, grass, or other produce of a forest; the right to prohibit the pasture of cattle in a forest; the right to cut, use, or remove any timber, trees, bamboos, branches, leaves, brush-wood, grass, or other produce of a forest; and the right of pasturing cattle in a forest.

PART II.

Of the Classification of Forests and of Forest Officers.

Forests to which these rules apply. 3. The provisions of these rules shall, in the manner hereinafter described, apply to the following classes of forests :

First.—Forests the property of Government in which the proprietary right of Government is absolute and unrestricted.

Second.—Forests the property of Government in which the proprietary right is limited by the existence of forest rights vested in other persons.

Third.—Forests which are the property of village or other communities, public institutions, religious establishments, private individuals, or other persons, but in which Government has forest rights.

Forests to which certain provisions only apply. 4. Except when otherwise provided in these rules, all forests other than those described in the last preceding rule shall be at the absolute and unrestricted disposal of the owners thereof.

Forest Officers. 5. The control, inspection, management, and protection of any forest to which these rules apply, or the control and management of any matters to which these rules relate, shall be vested in the following officers :—

I.—The Forest Officers.

II.—The Deputy Commissioners in charge of districts, and the subordinate revenue officers.

The Resident at Hyderabad may invest any person with all or any of the powers which may be exercised by a Forest Officer under these rules.

Deputy Conservator of Forests. 6. The designation of the chief Forest Officer in the Hyderabad Assigned Districts will, from time to time, be determined by the Government of India. At present he will be designated Deputy Conservator of Forests.

Forest Officers not to engage in other employments. 7. No Forest Officer shall engage, or be concerned in trading with timber, wood, or other forest produce, nor shall he have any interest, direct or indirect, in any lease of any forest, or any contract for working any forest, whether in British or foreign territory in India, nor shall he engage in any employment or office other than his duties under these rules, unless expressly permitted to do so in writing by the Resident.

Part III.

Of the Demarcation of Forests.

8. The Resident may, from time to time, notify the boundaries of any forest of the first

State forests.

and second classes described in Rule 3, and may declare the same to be a State forest. If in the judgment of the Resident such forest is not sufficiently demarcated by roads, streams, or other existing boundaries, or landmarks, the Resident shall, from time to time, cause the same to be marked out by metes and bounds, and the boundaries so fixed shall be notified in such manner as the Resident may direct.

9. The Resident may from time to time notify the boundaries of any forest of the second

Communal forests.

and third classes, described in Rule 3, and may assign all or any of the Government produce of such forest for the use of village or other communities, or for any other purpose. If in the judgment of the Resident such forest is not sufficiently demarcated by roads, streams or other existing boundaries, or landmarks, the Resident shall, from time to time, cause the same to be marked out by metes and bounds, and the boundaries so fixed shall be notified in such manner as the Resident may direct.

Demarcated forests.

10. Forests, the boundaries of which have been notified under Rule 8 or 9, shall be called "demarcated forests."

11. In the event of any difference or dispute arising as to the boundary line of any

Boundary disputes to be referred to the Deputy Conservator of Forests.

demarcated forest, such difference or dispute shall be referred to the Deputy Conservator of Forests, who shall endeavour to adjust the same by agreement with the parties interested.

12. If the matter in dispute cannot be arranged between the Deputy Conservator and

When to be referred to the Commissioner of the division.

the parties by mutual agreement, it shall be referred to the Deputy Commissioner, who himself, or an Assistant invested with full powers, shall decide the matter, subject to an appeal within one month to the Commissioner of the division, whose decision shall be final.

13. When the boundary line shall have been determined either by agreement or by decision

Deputy Conservator to appoint date for marking out the boundary.

of the Commissioner or Deputy Commissioner as aforesaid, the Deputy Conservator of Forests shall fix a day for marking out the boundary accordingly, of which public notice shall be duly given.

14. On the day appointed, the boundary shall be marked out by two or more Forest Officers

Record of boundary.

specially appointed by the Deputy Conservator of Forests in that behalf, and a record of the proceedings, together with a map showing the boundary lines determined as aforesaid, shall be prepared. When the boundary line has been determined by the Commissioner or Deputy Commissioner as aforesaid, the decision in writing of the Commissioner or Deputy Commissioner shall form part of the record. Such record and map, after being duly attested, shall be deposited in the office of the Deputy Conservator of Forests. An attested copy of the record and map shall also be deposited in the office of the Deputy Commissioner of the district or districts in which the forest or part thereof is situated; such map, record, and attested copy of the same shall be conclusive evidence of the position of the boundary line determined as aforesaid.

Part IV.

Of Forest Rights in Demarcated Forests.

15. From and after the date in which the boundaries of any demarcated forests shall have

Fresh forest rights not to accrue.

been notified, no fresh forest rights whatsoever shall accrue, or be acquired in such forest.

16. When the boundaries of any forest have been notified under Rules 8 and 9, the forest

Forest rights to be defined, determined, and recorded.

rights of the people living within the limits of the forest or outside in its vicinity, shall be defined and determined by such officers, and in such a manner as the Resident may from time to time direct: a record of the rights thus defined and determined shall be prepared and signed by the Deputy Conservator of Forests, and shall be deposited in his office, and an attested copy shall be deposited in the office of the Deputy Commissioner of the district or districts in which the forest or part of it is situated. No forest rights which have not been defined and determined and recorded in manner aforesaid shall be recognized or admitted.

17. In any demarcated forest, the Deputy Conservator of Forests may, from time to time,

Roads in demarcated forest.

in consultation with the district officer, and with the sanction of the Resident obtained through Commissioner of the division, determine what roads and pathways shall be authorized for public traffic, and may cause all other roads and pathways to be closed either permanently, or for a time only. Public notice of the closing of any existing road or pathway shall be duly given.

F

18. It shall be lawful for the Deputy Conservator of Forests, with the sanction of the Resident, from time to time, to close such portions of any demarcated forest as may be deemed necessary for the protection and improvement of the said forest.

Portions of demarcated forests may be closed.

The portions closed at any one time shall not exceed one-third of the total area of such forest.

Public notice of the situation of the portions closed shall be duly given and temporary marks shall be put up and maintained to indicate the boundaries of the portion so closed.

19. During such time as any portion of a forest shall be closed, no person having forest rights shall exercise the same within such portion, and no person shall cut or collect any timber, wood, grass, or other forest produce, or take or pasture any cattle therein, but the Deputy Conservator of Forests shall, if necessary, provide an equivalent for the rights so suspended.

Forest rights not to be exercised in any closed portion of a demarcated forest.

20. Except with the permission in writing of the Deputy Conservator of Forests, no forest rights in any demarcated forest, whether vested in communities, or other persons, shall be alienated by sale, lease, mortgage, or other contract, and no forest produce obtained by the exercise of such forest rights shall be sold or transferred to any person not entitled to the exercise of such forest rights.

Forest rights and produce obtained by the exercise of forest rights not to be alienated.

PART V.

Of the Regulation, Restriction, Commutation, and Extinction of Forest Rights in Demarcated Forests.

21. In any demarcated forest, the Resident may, when such may appear necessary, order the regulation, restriction, commutation, or extinction of any or all forest rights :

Regulation, restriction, commutation, and extinction of forest rights.

Provided that in all cases where any existing forest rights of communities or other persons are affected or abridged by such orders, fair and equitable compensation shall be given to such communities or other persons. Such compensation shall consist either in the grant of land, or of forest rights in other forests or lands, or in the payment of a sum of money, or in the abandonment of the rights of Government in any forest or other lands, or in the remission or reduction of any taxes, or of any other Government demand, or in any of these conjointly. The nature and amount of such compensation shall be determined in the manner hereinafter provided.

Proviso.

22. Whenever the regulation of any forest rights shall be ordered under the last preceding rule, such regulation shall consist in determining the following matters according to the nature of the right to be regulated :—

Matters to be determined when forest rights are regulated.

The quantity and description of timber or wood which shall be delivered to those entitled thereto annually or at stated seasons.

The time and place at which requisitions for such timber or wood shall be made, and the time, place, and mode of delivering the same.

The quantity and description of timber or wood which may be cut and removed by those entitled thereto, the extent and situation of the portions of forest where such cutting will be permitted, the seasons and time of the day when the cutting and removal of such timber or wood will be permitted, the lines for its removal, and the places where such timber or wood shall stop to be examined and passed.

The extent and situation of the portions of forest which shall be assigned annually, or at stated seasons, for the lopping of branches, the cutting of grass, and the collecting of other forest produce to those entitled to such rights.

The extent and situation of the portions of forest which shall be assigned every season to those entitled thereto for the pasturing of cattle, and the number of heads of cattle which may be sent into the forest.

The date on which the season for pasturing cattle shall commence and close.

23. Whenever the Resident may consider it necessary that any or all forest rights exercised in a demarcated forest shall be regulated, restricted, commuted, or extinguished, the Deputy Conservator of Forests shall prepare a statement of the existing forest rights, and frame proposals for regulating, restricting, commuting, or extinguishing the same, and for the compensation to be granted to any person entitled thereto.

Proposals to be made by Deputy Conservator.

24. The statement and proposals of Deputy Conservators of Forests shall be submitted to the Resident, and, if approved, shall be communicated to all parties interested. If the said parties shall consent thereto, or otherwise if any objection made thereto be adjusted by mutual agreement, such agreement being approved by the Resident, a record of the terms of such consent or agreement shall be reduced to writing, and be attested by the official signature of the Deputy Conservator of Forests. Such records shall be conclusive evidence of the terms of such agreement or consent. It shall be deposited in the office of the Deputy Conservator of Forests, and an attested copy of the same shall be deposited in the office of the Deputy Commissioner of the district or districts in which the forest or any part of it is situated.

Proposals to be communicated to persons interested.

25. If the proposals of the Deputy Conservator are not agreed to, or the objections
thereto cannot be adjusted by mutual agreement, the
compensation shall be determined according to the provisions of Act X of 1870 :

Procedure on disagreement with Deputy Conservator's proposals.

Provided that if in any case the decision shall award such amount of compensation that the
Deputy Conservator of Forests shall deem it better to
leave such forest rights unregulated, unrestricted, uncommuted, or unextinguished, than to give such compensation, it shall be lawful for him, with the
sanction of the Resident, to decline to give the compensation awarded as aforesaid, and in such
case the person entitled to such forest rights shall continue to enjoy the same rights without
such regulation or abridgment.

Proviso.

Part VI.

Of the Control over Demarcated Forests.

26. All demarcated forests shall be under the exclusive control of the Deputy Conservator
of Forests in Berar, and of the subordinate Forest Officers, except in the cases in which the Resident may deem
expedient to delegate the charge of any such forests to
a district officer.

Demarcated forests to be under the exclusive control of Forest Officers.

27. If any land under cultivation before the demarcation of the forest is included in the
demarcated limits, the cultivators' rights in the same
shall not be interfered with, excepting under the provisions of Act X of 1870.

Cultivators' rights in demarcated limits how to be dealt with.

28. No other land included in any demarcated forest,
&c., shall be alienated or leased out for a period exceeding
one year without the sanction of Resident.

Land in demarcated forests not to be alienated or leased for more than a year without sanction.

29. No person shall without permission carry or
kindle fire within the limits of any demarcated forest.

Fires within demarcated forests prohibited.

30. No persons except those authorized by special permission, or whose rights shall have been
admitted in accordance with Rule 16, shall enter any demarcated forest, or pass through the same except on roads
or pathways authorized in accordance with Rule 17, and no
person shall use any road or pathway which may have been closed in accordance with that rule.

Passage through demarcated forests, except on authorized roads, prohibited.

31. Within the limits of any demarcated forest, no persons save those authorized by special permission, or in accordance with Rule 16, shall
carry axes, saws, or other implements for cutting wood
or timber, except on roads and pathways authorized by
the Deputy Conservator of Forests.

Carriage of implements for cutting wood prohibited.

32. The grazing and passage of cattle without permission or authority given in accordance with Rule 16 is prohibited within the limits of any
demarcated forest. No cattle may pass on any except
authorized roads and foot-paths, and any cattle found trespassing within the limits of any demarcated forest may be seized and detained in the manner hereinafter described.

Grazing and passage of cattle prohibited.

33. Within the limits of any demarcated forest, no tree shall be marked, girdled, lopped,
felled, burnt, or removed ; nor shall any brush-wood, bamboos, grass, or other produce of such forest be cut,
burnt, collected, or removed without special permission
or authority granted in accordance with Rule 16.

Marking, girdling, lopping, felling, burning or removal of trees, brush-wood, grass, bamboos, &c., prohibited.

34. No land shall be cleared or prepared for cultivation, no dhya shall be made, and no
lime or charcoal shall be burnt within the limits of any
demarcated forest without special permission or without
authority granted in accordance with Rule 16.

Clearance of land for cultivation, dhya cultivation, and lime and charcoal burning prohibited.

Part VII.

Of the Control over Forests which are not demarcated.

35. All forests to which these rules apply, and the boundaries of which have not been
notified under Rule 8 or 9, shall ordinarily be under
the control of the Deputy Commissioner of the district
in which they are situated, subject to inspection, and
with the co-operation of such Forest Officers as the Resident may from time to time direct.
But the Resident may at any time place such forests under the exclusive control of Forest
Officers or other persons vested with the powers of Forest Officers in accordance with Rule 5.
Such forests shall be designated district forests.

Control over forests not demarcated by whom to be exercised : such forests to be designated district forests.

36. District forests shall consist of reserved and
unreserved.

What district forests shall consist of.

37. The Resident may, at any time on application of the Deputy Conservator of Forests
through Commissioner, declare that the waste land of
any particular villages of a district are to be reserved,
and a register of such reserved district forests in each
taluk shall be prepared and forwarded to the Deputy Conservator of Forests.

The Resident may at any time declare waste lands of any particular villages of a district to be reserved.

38. No land in any reserved district forest shall be alienated or leased out without previous inspection and the concurrence of the Deputy Conservator of Forests, and land thrown up by cultivators shall not be re-let without such concurrence of inspection.

Alienation or lease of land in reserved district forests prohibited.

39. Dhya cultivation, whether in fresh forest, or in old clearings, is prohibited in all district forests. In special cases permission to make dhyas may be given by the Deputy Commissioner, or by the Forest Officer in charge of the forest.

Dhya cultivation prohibited.

40. The following six kinds of trees shall be reserved in all district forests, and may not be felled, cut, burnt, lopped, or injured, nor may their leaves be gathered without the permission of the officer in charge of the forest :—

Reserved trees.

1.—Teak	...	...	...	(*Tectona grandis.*)
2.—Sheshum	...	...	...	(*Dalbergia latifolia.*)
3.—Beejasal	...	...	...	(*Pterocarpus Marsupium.*)
4.—Tinus	...	...	...	(*Dalbergia oojeinensis.*)
5.—Unjun	...	...	...	(*Hardwickia binata.*)
6.—Babool	...	...	...	(*Acacia arabica.*)

The Resident shall have power to revise or enlarge this list of reserved trees.

41. The Resident shall have power to sanction rules framed by Deputy Conservator of Forests, and submitted through Commissioner of division, for the felling, cutting, removal, and disposal of the reserved kinds of trees or their produce, and also for the general management of district forests, and to decide the terms on which pasture of cattle may be permitted, lime and charcoal may be burnt, and fruits and leaves, grass, gums, wax, lime-stone, and other forest produce may be collected and disposed of in any district forest, and to prescribe penalties for the non-observance of such rules, provided that such penalty shall not exceed a fine of Rs. 500, or in default, simple imprisonment for six months; and that the offences shall be classified as those for which the offender can be arrested without warrant, and those for which a warrant is necessary.

Resident's power in forest matters defined.

PART VIII.

Of the general Control over Lands and Forests, the property of Government or other Persons.

42. When such may appear necessary, the Resident may, by notification in the official Gazette, prohibit the clearing, felling, or burning of trees, bamboos, brush-wood, or grass in any forest or part of a forest when the maintenance of such forest appears necessary for the following reasons :—

Protection of forests for special purposes.

1st.—For the preservation of the soil on the tops, ridges, slopes, and in the valleys of mountain ranges.

2nd.—For the protection of the banks of mountain streams, rivers, and other waters.

3rd.—For the maintenance of a water-supply in springs and streams.

4th.—For the protection of any land against shifting and moving sands.

5th.—For the protection of roads, bridges, railways, and other lines of communication.

6th.—For the better preservation of the public health in the vicinity of such forest.

This rule shall be applicable to all forests enumerated in Rules 3 and 4, and the prohibition provided for in this rule shall be enforced, notwithstanding any forest rights or other privileges to the contrary.

43. The Resident may, when such may appear necessary, direct that possession shall be taken on behalf of Government of forests or lands the property of communities or other persons in which Government has forest rights, or in which Government has no forest rights. In such cases fair and equitable compensation shall be given to the owners of such land or forest, either by relinquishing the rights of Government in some other forest or land, or by the grant of land, or by the grant of forest rights in other forests or lands, or by the payment of a sum of money, or by the remission or reduction of taxes or any other, or any other Government demand, or by any of these conjointly.

Expropriation of land in which Government has or has not forest rights.

44. When the Resident shall direct that possession be taken of any forest under Rule 43, the Deputy Conservator of Forests shall draw up a statement of the situation and boundaries of the land or forest proposed to be taken, of the owners and others having forest rights or other beneficial interest in the land, of the nature and extent of the forest rights of Government in the land, and of the compensation proposed to be given, and shall submit the same to the Resident. Such statement and proposals, if approved, shall be communicated to all parties interested. If the said parties shall consent thereto, or otherwise if any objection made thereto be adjusted by mutual agreement, such agreement being approved by the Resident, a record of the terms of such consent or agreement shall be drawn up and compensation given accordingly.

Proposals for compensation.

45. If the proposals of the Deputy Conservator of Forests are not agreed to, or the objections thereto cannot be adjusted by mutual agreement, the compensation shall be settled in the manner laid down in Act X of 1870 : Provided always that, if in

Course to be adopted when proposals of Deputy Conservator of Forests are not agreed to.

any case the decision shall award such amount of compensation that the Deputy Conservator of Forests, acting with the sanction of the Resident, shall deem it better that possession be not taken of such forest, it shall be lawful for him to decline to give the compensation awarded as aforesaid, and in such case the forest shall continue in the occupancy of the owner, subject to the provisions of these rules.

PART IX.

Of the Control of Timber and other Forest Produce in Transit.

46. The Resident may from time to time, by notification in the official Gazette, prohibit the closing or obstruction for any purpose whatsoever of the channel or banks of any river, stream, or other water which is used for the transport of timber or forest produce, whenever such closing or obstruction is likely to interfere with the transport of such timber, wood, or other forest produce.

Streams not to be closed or blocked up.

47. The Local Administration may authorize the levying of dues or revenues on timber or other forest produce brought from any forests situated within the Hyderabad Assigned Districts, and may determine the rates of such duty. And the Local Administration shall have a lien on all such timber or other forest produce for the payment of such dues or revenues, and in the event of their not being paid and adjusted within a certain time to be fixed by the Resident, the timber may be confiscated.

Duty to be levied on timber.

48. The Deputy Conservator of Forests, with the Resident's sanction, shall from time to time determine and notify the stations at which all timber and forest produce in transit shall be stopped for the purpose of examining the same, and levying the duties and revenues lawfully payable thereon.

Stations at which duties are to be levied to be determined.

PART X.

Procedure of Forest Officers and others for the Prevention, Detection, and Punishment of Offences against these Rules.

49. It shall be the duty of every Police and Forest Officer to prevent, and he may interpose for the purpose of preventing the commission of any offence punishable under these rules.

Duties of Police and Forest Officers.

50. It shall be the duty of all communities or other persons who may have forest rights in any forest to which these rules apply, and of all persons who may have acquired by lease or otherwise the right of cutting, using, or removing timber, wood, and other forest produce, or of pasturing cattle in such forest, to assist to the utmost of their ability in preventing the occurrence of fires in the forest in which they have any right or interest as aforesaid. In the event of a fire occurring in such forest, it shall be the duty of all such communities or persons to aid in extinguishing the same, and to use due diligence in endeavouring to detect and bring to justice any person or persons who may be suspected of having been concerned in causing such fire.

Duties of persons who have forest rights.

It shall also be the duty of the communities or persons as aforesaid to prevent the commission of any offence punishable under these rules, and they shall use due diligence in endeavouring to detect and bring to justice any person or persons who may be suspected of having been concerned in the said offence.

51. Any Police or Forest Officer may seize and detain any cattle found straying or trespassing, or doing damage to any forest to which these rules apply; all cattle thus seized shall be dealt with according to the provisions of Act No. III of 1857, or such other Act relating to trespass by cattle as may be in force at the time in the Hyderabad Assigned Districts.

Power of seizing and detaining cattle.

52. The Commissioner of the district may from time to time establish cattle pounds in places where the Forest Officers may declare them necessary.

Cattle pounds by whom established.

53. Any Forest or Police Officer may arrest without warrant any one who in his presence is guilty of any of the offences specified in paragraphs 68, 70, 71, 72, 73, 74, and 78, hereafter described.

Arrest without warrant when effected.

54. For any other offence the Forest Officer, or Police Officer aforesaid, shall apply for a summons to the Magistrate having jurisdiction.

Magistrate's summons when to be applied for.

55. Every offence against these rules is bailable, and if sufficient bail is offered, the person arrested shall be released and bound over to appear before the Magistrate having jurisdiction.

Offences against these rules bailable.

56. In case of an offence being committed in which the Forest Officer or Police Officer aforesaid is not authorized to arrest the offender without a warrant, if the offender is not known to the officer, and it is suspected that the offender has not given a true address, the officer aforesaid may arrest and forward forthwith to the Magistrate or hold the defendant to bail.

Arrest of an offender not known to the officers, or who has not given a true address.

57. Whenever any person charged with an offence against these rules is arrested, he shall be forthwith, if not bailed, forwarded to the Magistrate having jurisdiction.

Offenders against these rules to be sent to Magistrate when arrested.

58. Every arrest made shall be reported to the Magistrate of the district when the arrest is made.

All arrests to be reported to the Magistrate.

59. All timber or other forest produce unlawfully removed, or attempted to be removed from any forest to which these rules apply, or from any station appointed by the Deputy Conservator of Forests under Rule 49, or from the custody of any Forest Officer, or dealt with or attempted to be dealt with contrary to rules framed under the provisions of Rule 41, or for which the dues and revenues lawfully payable thereon have not been adjusted, or for which a pass or other proof of ownership is not produced as prescribed by Rule 48; and all implements, tools, boats, and carts used in infringing any of the provisions of these rules, shall be liable to be seized by any Police or Forest Officer, and shall be dealt with in the manner provided in the next rule.

Seizure of forest produce or implements.

60. When any timber or other property shall be seized, any Magistrate having local jurisdiction may, upon information or otherwise, summon the person found in possession of such timber or other property, or suspected to have been concerned with the same, and upon his appearance, or in default thereof in his absence, shall examine into the cause of the seizure of such timber or other property; and in the case of timber or forest produce described in the last preceding rule, may direct the same to be restored to its lawful owner, or, no such owner appearing, to be confiscated and sold on account of Government; and in the case of tools, implements, boats, or carts used in infringing any provision of these rules, may direct the same to be confiscated and sold on account of Government. The sale of timber or other property confiscated under this rule shall be effected in such manner as the Resident may from time to time direct. Such confiscation may be enforced in addition to any fine or penalty imposed.

Confiscation of forest produce and implements.

61. In all cases where the confiscation and sale of such timber, forest produce, implements, tools, boats, and carts shall be adjudged as aforesaid, it shall be lawful for the Magistrate adjudging the same to award a portion of the proceeds of such sale, not exceeding one-half, to any person on whose information such property has been seized.

Portion of proceeds may be awarded to informer.

62. A Magistrate or Subordinate Magistrate of the 1st class shall have power to try and punish under these rules.

63. The provisions of the Criminal Code of Procedure shall be applicable to issue of processes, and the trial of all cases of infringement of provisions of these rules, and to all appeals from orders passed on such trials, and to the enforcement of penalties.

Provisions of the Criminal Code of Procedure made applicable to issue of processes.

64. When any confiscation or penalty shall be adjudged under these rules, the Resident may, within three months after final judgment, call for the proceedings of the case, and, if he shall see cause, may direct that the property seized, or any part thereof, be restored, and may remit the penalty or any part thereof, and direct that the offender be discharged.

Remission of penalties.

65. Any suit against any person for anything done, or purporting to have been done in pursuance of these rules, shall be commenced within three months after occurrence and not otherwise.

Limitation of suits.

Notice in writing of every such suit and of the cause thereof shall be given to the intended defendant one month at least before the commencement of the suit. The plaintiff shall not recover if tender of sufficient amends is made before suit, or if a sufficient sum of money is paid into court after suit brought by, or on behalf of, defendant.

If the suit has been commenced before tender of sufficient amends has been made, or a sufficient sum paid, then the defendant shall be liable to costs already incurred.

66. No charge for an offence against these rules shall be instituted except within six months after such offence has been discovered by a Forest or Police Officer.

Period within which charges to be brought.

67. A charge of an offence punishable under these rules, or under any other law for the time being in force, of which any Forest Officer whose appointment is notified in the official Gazette is accused as such Forest Officer, shall not be entertained against such Forest Officer without the sanction of the Resident.

Prosecution against Forest Officers.

PART XI.

The Offences punishable under these Rules are:

68. After a portion of a demarcated forest has been closed, any one found exercising forest rights, or cutting, collecting any timber, grass, wood, or other forest produce, or taking, or pasturing cattle therein.

Without warrant.

69. The sale or transfer without permission of forest produce obtained by the exercise of forest rights to any person not entitled to such rights.

Warrant required.

70. Kindling or carrying fire without permission within the limits of any demarcated forest. *Without warrant.*

71. Trespass by unauthorized persons in a demarcated forest-passage through such forests, except by authorized pathways and roads. *Without warrant.*

72. Trespass on a pathway or road closed under Section XVII. *Without warrant.*

73. Carrying within a demarcated forest by an unauthorized person of an axe, saw, or other implement for cutting timber and wood, except on roads and pathways authorized by Deputy Conservator of Forests. *Without warrant.*

74. Cutting, lopping, marking, girdling, felling, burning, or removing any tree, or burning, cutting, collecting, or removing any forest produce, brush-wood, grass, or bamboos, without authority, in any demarcated forests. *Without warrant.*

75. Clearing or preparing, without due authority, land for cultivation, burning lime or charcoal without proper authority in any demarcated forest. *Warrant required.*

76. Making a dhya in any State or district forest without special permission. *Warrant required.*

77. Cutting, lopping, marking, girdling, felling, burning, or removing any of the reserved kinds of timber in district forests, or collecting their leaves. *Warrant required.*

78. After a prohibition under Rule 42 has been duly notified, clearing, felling, burning trees, bamboos, brush-wood, or grass in any forest to which the prohibition extends. *Without warrant.*

79. Closing or obstructing any channel or bank of a river, after such a closing has been prohibited under Section 46. *A summons.*

80. Fraudulent attempt to evade payment of any dues or taxes lawfully due. *Summons.*

81. Any one convicted before a Magistrate of any of the above offences shall be liable to a fine not exceeding Rs. 500, and in default to a simple imprisonment for a term not exceeding six months.

82. Whosoever abets with in the meaning of the Indian Penal Code any offence punishable under any of these rules, shall be liable to the same punishment as is provided for such offence. *Abetment of offences. Warrant required or not, according to whether it is necessary for the offence abetted.*

83. Nothing in these rules shall be construed to prevent any person from being prosecuted under any other Regulation or Act for any offence punishable by these rules, or from being liable under any other Regulation or Act to any other or higher penalty or punishment than provided for such offence by these rules : Provided that no person shall be punished twice for the same offence. *Provisions of these rules not to prevent the operation of other laws.*

84. For wilful neglect of duty, or for negligent and careless discharge thereof, the Deputy Conservator of Forests may sentence any Forest Officer under his control, whose appointment is not notified in the Gazette, to pay a fine not exceeding one month's pay of such Forest Officer. *Forest Officers liable to punishment for neglect of duty.*

85. Any Police or Forest Officer who shall vexatiously and unnecessarily seize any moveable property of any person under the pretence of taking property liable to seizure or confiscation, or who shall vexatiously and unnecessarily arrest any person, or commit any other excess beyond what is required for the execution of his duty, shall be liable to fine not exceeding five hundred rupees, or in default to simple imprisonment for a term not exceeding six months. *Penalty for vexatious seizures and arrests.*

86. Any direct infraction of Rule 51, failure to perform the duties therein noted, will render the said persons or communities liable to a forfeiture of their forest rights.

PART XII.

Rules for the Grazing of Cattle in the Mailghât Hills.

I.—All parties desirous of grazing their cattle within the limits of the Mailghât Hills can obtain pass-notes at the under-mentioned offices, on payment of the prescribed fee. No pass can be taken out for a shorter period than one year :— *Rule for registering cattle.*

II.—Extra Assistant Commissioner, Argaon. *Offices from which pass-notes can be obtained.*

Tahseeldar,	Oomrawuttee.
Do.,	Chandore.
Do.,	Moortizapore.

Tahseeldar, Durriapore.
Do., Ellichpore.
Do., Moresee.
Do., Akolah.
Do., Akote.
Do., Julgaon.
Naib Tahseeldar, Chiekuldah.
Deputy Conservator of Forests.

Established fees.

III.—Buffaloes, including young, per 100, Rs. 12. Cows, including young, per 100, Rs. 5.

Privilege granted to inhabitants of Mailghât.

IV.—The inhabitants of Mailghât will be allowed free grazing per plough, *i. e.*, one pair of bullocks to each plough in *use.*

Police to assist in seizure of stray cattle.

V.—It shall be the duty of village and other Police Officers, when called upon, to assist in the seizure of cattle trespassing.

Graziers always to have pass-notes with them.

VI.—Cattle graziers must be prepared to produce their passes when called for by Forest Officers ; failing to do so, from whatsoever cause, they shall be taken before the nearest officer mentioned in Rule II, who shall issue a pass, the cost of which, calculated at double the ordinary and prescribed rates, shall be recovered forthwith from the owner of the cattle seized. This rule will likewise hold good in the case of herdsmen grazing cattle in excess of the number entered in the pass.

Separate passes necessary.

VII.—Each cattle owner must take out a separate pass-note in his own name for the number of cattle he intends registering ; but one grazier may graze the cattle of many owners.

Cattle found grazing without herdsmen.

VIII.—All cattle found grazing, not in charge of a herdsman, will be regarded as stray cattle, and conveyed to the nearest established pound with as little delay as possible, and there detained and disposed of according to Act III of 1857.

Fee for copy of pass-notes.

IX.—Any person losing his pass-note can obtain a copy at any of the offices named in Rule II, on payment of a fee of eight annas.

Pass must be returned.

X.—At the expiration of the period for which a pass has been obtained, or at any period a person may wish to withdraw his cattle from forest limits, the pass must be returned to one of the officers noted in Rule II, whose duty it will be to forward the same to the Deputy Conservator. A breach of this rule will render the offender liable to a fine of one rupee.

XI.—All officers issuing passes will, at the expiration of the month, be required to report to the Deputy Conservator the number of passes issued, balance in hand, and amount of fees collected.

XI.—MYSORE

DRAFT OF MYSORE FOREST RULES.

CHAPTER I.

General.

1. The administration of the forests will be vested in the following officers in the manner hereinafter described :—

Officers appointed for their administration.

I.—The Conservator of Forests, his assistants, and the subordinate Forest Officers.

II.—The Deputy Commissioners in charge of districts, and the subordinate Revenue Officers.

2. It will also be the duty of all Police Officers to watch over the observance of these rules, and to afford every assistance to the Forest Officers in the exercise of their duties.

Duties of Police Officers under these rules.

3. State forest means any forest reserve and plantation the property of the State, which has been demarcated and notified as a State forest under the authority of the Chief Commissioner, as prescribed in Rules 10, 11.

State forests.

4. The Chief Commissioner may from time to time direct village forests to
Village forests. be formed from such portions of waste or forest land as have not been included within State forests. The provisions regarding the demarcation and management of such forests will be the same as those laid down for the State forests, with such modifications as the Chief Commissioner may from time to time direct.

5. District forests shall mean such unoccupied Government land, outside
District forests. the limits of State and village forests, as may be declared to be district forests by competent authority.

Control of State forests. 6. State forests are managed exclusively by the Conservator of Forests and his assistants.

7. Village and district forests shall be managed by the Revenue Officers
Control of village and district forests. with the co-operation in certain matters of the Conservator of Forests and his assistants.
Where it appears necessary, the management of district or village forests in any locality may be handed over, under the authority of the Chief Commissioner, to the Conservator of Forests.

8. The supervision over ever-green forests used for the cultivation of the
Control of lands over which Government has forest rights. peppervine and toddy-palm, called kāns, and other lands in the occupancy of private individuals, in which Government possesses certain rights, is exercised by the Revenue Officers.

9. In these rules " cattle " shall be held to include, besides horned cattle,
Definition of " cattle." elephants, camels, horses, asses, mules, sheep, goats, and swine.

CHAPTER II.

Of State Forests.

10. The boundaries of State forests will, wherever they do not run along a
State forests how demarcated. road or stream, or other well-defined line, be demarcated by cleared boundary paths and permanent boundary-marks, or in such manner as the Chief Commissioner may direct. The boundary lines of State forests and the boundary-marks will be entered on maps which will be prepared in triplicate ; one copy to be in charge of the Conservator of Forests, one to remain with the Forest Officer in charge of the forest division, the third to be deposited in the office of the chief Revenue Officer of the district. The Chief Commissioner may dispense with the preparation of the map showing the boundary lines.

11. Proclamatoin of the formation of State forests and their boundaries
Proclamation. will be publicly made in the talook where they are situated, and will be published in the *Mysore Gazette.*

12. In State forests no land shall be alienated or leased, without the
Alienation of State forests. orders of the Chief Commissioner of Mysore.

Acts prohibited. 13. The following acts are prohibited in State forests :—

(1) Setting fire to the grass or forest, or kindling any fire in it or in the vicinity thereof, without effectually preventing its spread into the forest.
(2) Burning lime or charcoal without permission.
(3) Trespass by men or cattle off the authorized roads and pathways.
(4) Grazing or pasturing of cattle, except with the permission of the Conservator of Forests.
(5) Felling, girdling, cutting or lopping, marking, burning, stripping off bark or leaves, tapping for gum or resin, or otherwise injuring any trees, shrubs, or bamboos, except with the permission of the Conservator of Forests.
(6) Removal of dead leaves, turf, or the surface of the soil, cutting grass, collecting fruits, honey, wax, bark, gum, lac, or any kind of forest produce, without the permission of the Conservator of Forests.

(7) Temporary clearings called Kumri, or every other form of cultivation without the permission in writing of the Conservator
of Forests.

14. Existing roads or pathways through the State forests may be used
as far as is compatible with the conservancy
Roads and pathways.
of the forests; but the Conservator of Forests,
with the concurrence of the Commissioner of the division, may close any
existing road or pathway through any State forest, whenever he may deem
it requisite to do so. Public notice of the closing of such road shall be
given in the talook or talooks where the forest is situated, and some other
convenient road shall be provided in the place of the road thus closed.

15. The State forests may be entered by persons in pursuit of wild
beasts; but such persons shall not take with
Hunting permitted.
them, within the limits of the State forests,
except with the permission of a Forest Officer, any axes, knives, nor any
carts, or other vehicles, pack-bullocks, or ponies.

16. Cattle found straying in a State forest may be pounded, and may
be redeemed on payment of a sum of money
Cattle trespass.
according to a scale of rates to be laid
down from time to time by the Chief Commissioner of Mysore, and in default of
payment of such sum of money within such a time as the Chief Commissioner may fix, the cattle shall be sold on behalf of Government. It
shall be lawful for the officer selling such cattle to award a portion of the
proceeds of such sale, not exceeding one-half, to any person on whose information such cattle were seized. All such fines shall be credited to the Forest
Department.

17. In the State forests in the Mal-nand, the reed known as the
woutlé is free to all, and may be cut and re-
Reeds are free.
moved without restriction.

CHAPTER III.

Of District Forests.

18. District forests will be in charge of Shaikdars, excepting those
which, in accordance with Rule 7, may be
Control.
placed in charge of the Conservator of Forests.
Each Shaikdar will be responsible for the due protection of the forests
under his care, and it will be his duty immediately to report any breach of
these rules to the Amildar of his talook, and generally to prevent all injury
to the forests. Whenever the district forests in any district or talook are
placed under the control of the Conservator of Forests, in accordance with
Rule 7, the same shall be notified in the *Mysore Gazette,* and shall otherwise be made known in the chief towns and villages of the district or talook
in such manner as the Chief Commissioner may from time to time direct.

19. District forests will be subject to inspection and periodical report by
the officers of the Forest Department. These
Departmental inspection.
officers may also, in communication with
the Revenue Officer in charge of such forests, undertake any operation connected with planting, cutting, thinning, or selling timber in those district
forests which, in the opinion of the Conservator of Forests, may require their
special attention.

20. In district forests the management of which has been made over to
the Forest Department, land shall not be
Grant of land.
given away without previous inspection by,
and the concurrence of, an officer of the Forest Department duly authorized
to inspect and report upon such land.

21. Within fifty yards of the banks of a hill stream or any of its
feeders, or within a radius of fifty yards
Forest on hill streams and springs to be preserved.
from any spring, or within fifty yards of
any road, no trees, shrubs, bamboos, or jungle are to be cut in any district
forest except by the special direction of the Conservator of Forests. The
forest is not to be burnt, and nothing is to be done that may in any way
interfere with the growth of trees, brush-wood, or bamboos in such places.

In the case of roads, the Conservator may delegate his authority under this rule to the officer in charge of the road.

22. Kumri cultivation, whether in fresh forest or in old clearing, is prohibited in all district forests, special cases excepted, in which it may be permitted by the Chief Commissioner.

Kumri cultivation prohibited.

23. In district forests the following nine kinds of trees will be reserved, and may not be felled, cut, marked, or lopped, without the written authority of the Conservator of Forests or his assistants :—

Reserved trees.

(1) Sandal-wood—*Santalum album.*
(2) Teak—*Tectona grandis.*
(3) Blackwood—*Dalbergia latifolia.*
(4) Hone—*Pterocarpus Marsupium.*
(5) Lac, jalári—*Vatica laccifera.*
(6) Nundi—*Lagerstrœmia microcarpa.*
(7) Wild jack, hesswa, heb-halasu—*Artocarpus hirsutus.*
(8) Poon—*Calophyllum elatum.*
(9) Káráchi, kummar, arsina—*Hardwickia binata.*

Sandal-wood, teak, and poon will be sold by the orders of the Conservator, and under special conditions. The Chief Commissioner shall be at liberty from time to time to make such additions to or reductions in the number of reserved kinds of trees as he may deem fit.

24. In district forests firewood is free to all, with the exceptions noted in Rules 25–28: Provided always that none but unreserved matured trees and shrubs are cut. Dry sticks and branches of any kind lying unmarked in the forests may be taken. No reserved trees, whether saplings or matured or stunted trees, may be felled or lopped for firewood.

Firewood,

25. In places where it shall be found necessary, the Commissioner of the division, under the authority of the Chief Commissioner of Mysore, may introduce rules entirely prohibiting the felling of firewood within certain limits, or place such restrictions on the felling of firewood, as regards manner, time, place, quantity, and persons, as may appear requisite for the conservancy of any forest or jungle. He shall also be at liberty under the same authority to fix rates of seigniorage to be paid for such firewood, or order that payment be made in any other way that may appear to be the best.

Restrictions on cutting firewood.

26. Where it may appear to be necessary for the preservation of the forests, the Commissioner of the division, under the authority of the Chief Commissioner of Mysore, may place restrictions on the felling of wood for the burning of bricks or lime, for the making of charcoal, or for smelting iron ore. He shall also have power to limit the number of kilns for burning lime or charcoal, and the number of furnaces for smelting ore, in any district forest. He may cause all such kilns or furnaces to be registered, as also the persons employed in such work. He shall also have the power to fix the amount of seigniorage to be paid annually or otherwise for each kiln or furnace.

Wood for burning bricks and making of charcoal.

27. In any district forest the management of which under Rule 7 has been handed over to the Forest Department, the Conservator of Forests shall, under the authority of the Chief Commissioner of Mysore, have the same power of regulation, restriction, registration, and of fixing the amount of seigniorage to be paid, as by Rules 25 and 26 are conferred upon the Commissioner of the division.

Powers of Conservator of Forests in this matter.

28. All orders regarding such restrictions on the felling of firewood or for the registration of kilns and furnaces, and the rates of seigniorage fixed, shall from time to time be notified in the local Gazette under the orders of the Chief Commissioner of Mysore.

Orders to be notified in the Gazette.

29. In those talooks where, under the orders of the Chief Commissioner,
depôts for the supply of wood and bamboos
From Government depôts. have been or may be opened, the system of
licenses provided in Rules 35 to 38 shall be abolished, and all persons requiring
wood or bamboos, the property of Government, must buy such wood and
bamboos at the Government depôts. The place of such depôts, and the rates
to be paid for the wood and bamboos, shall be notified in the local Gazette
under the orders of the Chief Commissioner of Mysore, who may make such
changes in the rates and location of depôts as may from time to time appear to
be required.

30. All persons requiring unreserved wood or bamboos, the property of
Government in talooks where Government
License system. depôts have not been established, shall apply
to the chief Revenue Officer of the district for the needful license or permission;
such license or permission will not, however, be granted until the intending
purchaser has ascertained that the wood can be sold to him, and until
he produces a receipt showing that he has paid into the district or talook trea-
sury the full amount of duty. The rates shall be fixed for each district, and
published from time to time by the Conservator of Forests, under the authority
of the Chief Commissioner of Mysore.

31. The grazing of cattle, and the collection of gums, resin, lac, bees-wax,
Pasture and collection of minor forest pro- and other minor forest produce in district
duce. forests will be regulated in such manner as
the Chief Commissioner may from time to time direct.

32. Any licenses or permissions granted to traders, contractors, or others
for the felling or collection and removal of
Conditions of licenses. timber or any other forest produce, shall
contain such conditions regarding time, route, or method of removal, protection
of the forest against fire, and other damage, as may be considered necessary by
the officer granting the license.

33. Every license must be returned by the holder on expiration of the
time for which it was granted. All passes
Licenses to be returned. which may be issued by duly constituted
authority for the removal of any wood or other forest produce must also be
returned on the expiration of the time stated in the pass.

34. In case of fire or on any other special occasion, the Conservator may,
Grants of wood and bamboos free or at re- with the Chief Commissioner's sanction, give
duced rates. any timber, wood, or bamboo from any forest
or depôt without payment, or at reduced rates.

35. In the district forests the following articles of forest produce will be
free to the land-cultivating ryot for his own
Privileges granted to cultivating ryots. use, but not for sale or transfer:—

1st.—Wood for agricultural implements, but not for carts, from
unreserved trees only.

2nd.—Unreserved wood, thorns, and bamboos for fences, stack floors,
and cattle pens.

3rd.—Branches of unreserved trees for manure and litter.

4th.—Grass for thatching.

36. Should the collection of these articles in any locality be so large as to
render such a measure necessary, the chie*
Restrictions when to be imposed. Revenue Officer of the district may prohibit the
promiscuous cutting and collection of any or all of these articles, and limit the
quantity to be cut or collected annually by each cultivating ryot, and issue such
orders regarding the localities where and the season when such articles may be
cut and collected as may appear necessary.

37. The Chief Commissioner of Mysore may, if it shall happen to be
necessary either for the purpose of preserving
Privileges when withdrawn. any forest, or because of the ryots of any
locality having abused the privileges granted them, withdraw the whole or any
part of the privileges detailed in Rule 35 in any locality.

38. The Chief Commissioner may from time to time exempt certain talooks
Talooks exempted from ordinary operation or portions of talooks from the ordinary opera-
of forest regulations. tion of the forest rules, and may permit un-

reserved wood and bamboos to be taken free by the agricultural population for building their houses and carts. In talooks so exempted such permission shall be given to cultivating ryots alone. As regards traders or any other person not a cultivating ryot, the forest rules shall be considered in full force throughout the whole province of Mysore. The Chief Commissioner may, at any time, remove from the list of exempted talooks, or exempted localities, any talook or locality which it may be necessary so to remove, and re-introduce the full action of the forest rules in such talooks or localities.

CHAPTER IV.

Of the State Forest Rights on lands in the occupancy of other persons and on Inam lands.

39. On all lands throughout Mysore, excepting those in respect of which
Sandal-wood. the right is expressly alienated, the sandal-wood tree is the property of the State. The forest rights of the State on such lands with regard to sandal-wood will be exercised under such rules as the Chief Commissioner may from time to time lay down.

40. Trees of the reserved kinds referred to in Rule 23 standing on lands
State forest rights in lands taken up under settlement rules. taken up by private individuals under the operation of the survey and settlement rules, shall be the property of the State for one full year from the date of such lands having been taken up by the occupants. During that period of time the Government shall have the right of removing all trees of the reserved kinds standing on the land. All such trees, with the exception of sandal-wood, left standing on such lands after the close of one year from the date of occupancy, shall lapse to the occupant of the land. Sandal-wood trees growing, or that may thereafter grow up on such lands, shall always be the property of the State, and shall not lapse to the occupant at any time: Provided that occupiers of land are authorized to remove from the same all sandal-wood seedlings of less than one yard in height, where such seedlings interfere with cultivation.

41. In talooks in which the survey and settlement have not yet been
State forest rights in unsurveyed and un-settled talooks. completed, no seigniorage is chargeable on the felling of any bamboos or trees of any kind (excepting sandal-wood and teak) standing on lands in the occupancy of private individuals, when such trees or bamboos have been planted by the present holder, or by his own immediate ancestors, or by the former occupant of the land from whom the present holder may have legally purchased the patta rights.

Any individual wishing to sell such trees or bamboos shall apply for, and obtain from the duly constituted authority, a free pass for the removal of the timber.

42. All inamdars who, by the terms of their sanads, are entitled to the
Forest rights of inamdars. timber and sandal-wood in their inam villages, are at liberty to fell and sell without previous reference any such timber (excepting sandal-wood) of the reserved or unreserved kinds. The Forest Department will, under the orders of the Chief Commissioner, fell and sell all sandal-wood growing in such inam lands, on behalf of the inamdars and under certain conditions.

43. Inamdars whose sanads do not make special mention of the holders
Forest rights of inamdars not entitled to timber and sandal-wood. being entitled to the timber and sandal-wood growing on their inam lands, will be allowed to fell and sell all unreserved wood and the reserved kinds of wood (with the exception of teak and sandal) growing on such lands.

Free passes. 44. All inamdars are required to obtain free passes for the removal of felled wood when sold.

45. Fruit-trees in public topes or in Government land not expressly relin-
Trees in topes. quished in the settlement effected by the Revenue Survey Department, and which do not fall within the scope of the forest regulations, cannot be cut down or lopped without sanction of the Chief Commissioner.

CHAPTER V.

Of the transport of wood and other forest produce.

46. All persons importing any timber or other forest products from Madras into Mysore on roads running through the State forests of the Mysore district of the Ashtagram division shall confine themselves to such roads as may from time to time be notified by the Chief Commissioner and marked by sign-posts at their entrance into and exit from the forests.

47. All drift and unclaimed timber and bamboos, and other forest produce within the limits of the Mysore territory, will
Drift and unclaimed timber. be considered the property of Government unless proof of ownership be given as hereinafter provided. All such forest produce shall be collected at such stations as the Conservator of Forests may from time to time direct, and notices shall from time to time be published stating the number and description of pieces of drift timber and bamboos and other forest produce collected at such stations.

48. Not less than two months will be allowed for the reception of claims to the ownership of drift and unclaimed tim-
Notice inviting claimants. ber, bamboos, and forest produce, after which no claims to such ownership will be received, and if no person has established his title to the said forest produce, the same will be sold on account of Government.

49. All such claims will be settled by the Conservator, or by such officer as he may authorize : Provided, however, that
Claims by whom settled. he shall be at liberty to decline arbitrating between rival claimants regarding such timber or bamboos, and in case he may see fit to do so, refer claimants to the civil courts.

50. Forest produce awarded to claimants must be redeemed by the payment of salvage and other expenses which may
Claimed timber how redeemed. have been incurred on account of such timber.

51. All timber marks used by traders in Mysore shall be registered in the office of the officer in charge of the forest divi-
Timber marks used by traders. sion, and each trader shall also deposit at that office a distinct and clear impression of all marks used by him, and also certified copies of all other private marks which he may use for the purpose of marking his wood in pursuit of his trade as a dealer in wood.

CHAPTER VI.

Of the prevention and punishment of offences against these rules.

52. For every breach of these rules, or omission contrary to the provisions thereof, the offender shall be liable, on
Penalties. conviction before a Magistrate having jurisdiction in the case, to fine not exceeding Rs. 500, or, in default of payment, to such imprisonment as is provided in Section 67 of the Indian Penal Code. Persons found off the authorized roads or pathways within the limits of any State forest, and owners of cattle straying in the forests, shall be liable to half the penalty provided in this rule.

53. Nothing in these rules shall be construed to prevent any person from being prosecuted under any other law for any act or omission which constitutes an offence against these rules, or from being liable under such other law to any higher punishment or penalty than that provided by these rules : Provided that no person shall be punished twice for the same offence.

54. Any axes, knives, carts, boats, or other tools, vehicles, or implements,
Tools, timber, and other articles may be confiscated. as also all cattle used in an act which constitutes an offence against these regulations, and all timber, wood or other forest produce which has been marked or obtained in a manner contrary to these regulations, whether entire or cut up or sawn up, and whether found within or outside the limits of the forest, and all timber, wood, or other forest produce, in transit by land or water, which is not covered by a pass as required by these rules, may be seized by any officer of the Forest Department or Police Officer, and such timber, tools, vehicles, implements, cattle, may be confiscated or released on payment of a fine by the orders

of the Magistrate of the district, or of any Magistrate specially empowered by the Chief Commissioner to exercise jurisdiction under these rules.

55. It is the duty of the officers and subordinates of the Forest Depart-

Duties of Forest, Revenue, and Police Officers, with regard to these rules. ment, and of all Revenue and Police Officers, to see that these rules are not violated, and should they in any case be infringed, to report the same to the nearest Magistrate or Forest Officer in charge of the forest division, and it shall be lawful for any Forest or Police Officer to take into custody, without a warrant, any person whom he may detect in any act constituting a breach of these rules, or any person who hinders or obstructs him in the discharge of his duties under these rules, and the person apprehended shall be brought before a Magistrate with the least possible delay.

XII.—DRAFT FOREST BILL FRAMED BY MR. BADEN-POWELL.

(*N. B.—The following Bill is prepared only in the form of " instructions to draft," without any attempt to meet the requirements of accurate legal wording, except in cases where sections from other Acts have been adopted.*)

WHEREAS it is expedient to provide—(a) for the better management and preservation of forests in British India; and (b) to regulate and protect the transport of timber and other produce of forests, whether by land or by water, through British territory, it is enacted as follows:—

PRELIMINARY.

Date of coming into operation. 1. This Act shall come into operation on the and may be cited as the " Forest Act, 1874."

Repeal of enactments. 2. The Acts* specified in the schedule hereto annexed are repealed, &c.; and all forest rules made under any of the said Acts are repealed also.

Extent of Act. 3. The Act extends to British India, except the territories of Madras and Bombay, and may be extended thereto by notification in official Gazette.

Definitions. 4. In this Act, unless there be something repugnant in the subject or context,—

" Forest" shall mean any tract of land producing, naturally or by cultivation, trees, brushwood, or bamboos, and shall include such other lands as may lawfully be declared to be " forest" for the purposes of the Act by public notification in the vicinity.

" Cattle" shall, besides horned cattle, include elephants, camels, horses, asses, mules, sheep, goats, and swine.

A mark placed upon a standing tree by Government, to indicate the property of Government therein or otherwise, shall be deemed to be a property-mark within the meaning of the Indian Penal Code.

" Forest right " shall mean—

(a).—the right of ownership in trees, timber, bamboos, brush-wood, grass, or other produce of a forest;
(b).—the right to cut, use, or remove any timber, trees, branches, leaves, or any bamboos, brush-wood, grass, or other produce of a forest or of the soil thereof;
(c).—the right to prohibit such cutting, user, or removal;
(d).—the right of pasturing cattle in a forest;
(e).—the right of prohibiting such pasturing;
(f).—the right of way through a forest;
(g).—the right of hunting, shooting, or fishing in a forest.

" Forest privilege" means any user of a forest or the produce thereof (as included in the definition of forest right), when such is exercised by permission and at the pleasure of the Local Government and not as of right.

" Government timber depôt" shall mean any place which the Local Government shall from time to time appoint for the storing of wood, timber, bamboos, or other forest produce, for the stoppage and examination of wood, timber, bamboos, or other forest produce in transit, or for the

* Note.—Section 2.—The rules legalized by the " Punjab Laws Act," which are a special thing, and contain a provision about fires (which is useful in Punjab, and would not be applicable elsewhere) should not be repealed.

B. H. B. P.

detention of the same for the levy or payment of duty, royalty, or otherwise, or for the storing of drift and other timber, bamboos, and other forest produce lawfully collected under this Act, or rules made under it.

River.   "River" includes stream, canal, creek, or other channel, natural or artificial.

5. The control, inspection, management, and protection of any forest to which this Act applies, or the control and management of any matters to which this Act relates, shall be vested in officers to be called Forest Officers. The Local Government may invest any person with all or any of the powers which may be exercised by a Forest Officer under this Act.

Forest Officers.

6. The chief Forest Officer in any presidency, province, or place, or part of a presidency, province, or place, in which this Act shall take effect, shall be styled "Conservator of Forests." The powers of the Conservator of Forests under this Act may be exercised by such forest or other officer as the Local Government may from time to time authorize in that behalf.

Conservator of Forests.

[7. No Forest Officer shall engage or be concerned in trading with timber, wood, or other forest produce, or in any agency business, nor shall he have any interest, direct or indirect, in any lease of any forest or any contract for working any forest, whether in British or foreign territory in India ; nor shall he engage in any employment or office other than his duties under this Act, unless expressly permitted to do so in writing by the Local Government.]

Forest Officers not to engage in other employment.

N. B.—If this clause is retained, it may be needed also to include the prohibition of Forest Officers owning land (not being sites of houses or buildings).

PART I.

OF THE ORGANIZATION OF FORESTS, AND THEIR PROTECTION AND MANAGEMENT.

CHAPTER I.

Forests to which the Act applies.

8. Except when otherwise expressly provided, the provisions of Part I of this Act shall apply to the following classes of forest only, that is to say :—

Classification of forests.

  1st.—Forests, the property of Government, in which the proprietary right of Government is absolute and unencumbered.

  2nd.—Forests, the property of Government, in which the proprietary right is limited or encumbered by the existence of other forest rights.

  3rd.—Forests which are the property of village or other communities, public institutes, religious establishments, or other persons, in which Government has forest rights.

9. (*a*) Any forest of the 1st or 2nd class, described in Section 8, (*b*) or any forest belonging to an estate which is managed by the Court of Wards, so long as it remains subject to such Court, (*c*) or any private or communal forest which the owner thereof desires to place under management of the Government (subject to such conditions regarding the payment of expenses and the termination of the management as the Government and the said owner may agree to), may be constituted a 'special reserve' or an 'ordinary reserve' by order of the Local Government.

Reserved forests.

10. A "special reserve" is a forest set apart for the production and supply of timber and other exceptionally valuable forest material, and consequently subject to strict preservation.

Special reserve.

In it no forest privilege shall be granted, except in special cases, where the absolute exclusion of such privileges would press hardly upon the people.

11. An "ordinary reserve" is a forest set apart for the supply and production of ordinary forest produce, and is subjected to general conservancy. Forest privileges may be granted in any ordinary reserve.

Ordinary reserve.

12. Whenever it is determined to constitute a reserve of either kind, the Local Government shall appoint a District or Settlement Officer and a Forest Officer, conjointly, to constitute such reserve, subject to the confirmation of the Local Government as hereinafter provided.

Constitution of reserves.

The officers charged with the constitution of such reserves shall, in making their proposals, consider both the convenience and well-being of the people resident in or in the vicinity of the forest, and also the forest requirements of the case, namely, the production of material and the utility of the forest in preserving the soil, the water-supply in springs and rivers, and the climate generally.

The constitution of reserves shall consist :—

(1.)—In the determination of the boundaries and the demarcation of the forest, as hereinafter provided.

(2.)—The settlement, definition, and record of all forest rights in such forest, in the manner provided by this Act.

(3.)—The settlement, definition, and record of all forest privileges which the officers charged with the constitution of the reserves deem it necessary to allow for the convenience and well-being of the residents in or in the vicinity of the forest.

13. If, at any stage of the constitution of the reserve, or in submitting proposals therea *Difference of opinion between officers in charge.* for, a difference of opinion arises between the District o- Settlement Officer and the Forest Officer, such differencr shall be decided by the chief revenue authority of the province, whose decision shall be final.

CHAPTER II.

Demarcation and boundaries.

14. In every special reserve, if the boundaries are not clearly and unmistakeably indicated *Demarcation of special reserves.* by roads, rivers, or other existing boundaries or land- marks, they shall be marked out by permanent marks in such manner as the Local Government may direct.

15. In every ordinary reserve, the boundaries shall ordinarily be fixed according to natur- *Demarcation of ordinary reserves.* al land-marks; but the officers whose duty it· is to lay down the boundaries may mark out the boundaries by such other marks as in any case they may consider it expedient.

16. In the event of any dispute arising as to the boundary of any such reserve, where it *Disputed boundary.* touches any other estate (not being Government proper- ty), the Civil or Settlement Officer appointed to lay down the boundary shall refer the dispute to arbitration, as hereafter provided.

17. When the boundaries have been determined, either by consent of the parties, by *Record of boundaries.* arbitration, or otherwise, a record of the boundaries, and, where possible, a map showing the boundary lines and marks shall be prepared.

18. When the boundary is by decision of arbitrators, the award shall form part of the *Award of arbitrators to form part of record.* record. Such record and map shall be deposited in the office of the Conservator of Forests, and a copy of it shall be deposited in the office of the Collector or other chief revenue authority of the district within whose jurisdiction the boundary line or any part of it lies.

19. In referring a disputed boundary to arbitrators, the officer making the reference *Reference to arbitration.* shall specify in the order of reference the precise matter submitted to the arbitrators, and such period as they may think reasonable for the delivery of the award, and may from time to time extend that period.

20. The Conservator of Forests and the other parties interested may each nominate *Appointment of arbitrators.* either one or two arbitrators, provided that each party shall nominate the same number. A third or fifth arbitrator, as the case may be, shall be appointed by the Civil or Settlement Officer making the reference.

21. If the parties aforesaid refuse to nominate arbitra- *Refusal to appointment.* tors, the Civil or Settlement Officer shall nominate three or five arbitrators, as he thinks fit.

22. Every officer making a reference under this chapter may, on good cause shown, *Power to excuse arbitrator from serving, and to call for nomination of substitute.* excuse any person from serving as an arbitrator, and may call on the party who nominated such person to nominate another in the place of the person so ex- cused.

23. If an arbitrator die, desire to be discharged, or *Nomination of new arbitrator in place of one dying or failing to act.* refuse or become incapable to act, the party who nomi- nated him shall nominate another person in his place.

24. If in any case any party fail for a week to nominate *Nomination by Civil Office when parties fail.* in manner aforesaid, the officer making the reference shall appoint some person to act as arbitrator.

25. The arbitrators shall determine and award concerning the matters referred to them *Award.* for arbitration ; and the parties disputing, and all per- sons claiming through them, respectively, shall abide by and perform the award of the arbitrators.

26. If the arbitrators require the presence of the parties or any other persons whose *Summoning parties to give evidence.* evidence may be necessary, they shall apply to the officer making the reference, who shall summon such parties or persons ;

and all such parties or persons shall be bound to attend, either in person or by agent, as the arbitrators may require, and to state the truth, and to produce such documents and other things as may be required before the arbitrators.

Obligation of persons summoned.

27. The award shall be made in writing under the hands of the arbitrators, and shall be submitted by them to the officer making the reference, who shall cause a notice to be served on the parties to attend and hear the award.

Preparation and submission of award.

28. The officer making the reference may remit the award, or any of the matters referred to arbitration, to the reconsideration of the same arbitrators—

In what cases award or subject of arbitration may be remitted to arbitrators.

(a) if the award has left undetermined some of the matters referred to arbitration, or if it determine matters not referred to arbitration ;

(b) if the award is so indefinite as to be incapable of execution ;

(c) if an objection to the legality of the award is apparent upon the face of the award.

29. No award shall be liable to be set aside, except on the ground of corruption or misconduct of all or any of the arbitrators.

Grounds on which award may be set aside.

Any application to set aside an award shall be made within ten days after the day appointed for hearing the award.

Application to set aside.

30. If the officers making the reference do not see cause to remit the award or any of the matters referred to arbitration for reconsideration in the manner aforesaid,

Decision according to award.

and if no application has been made to set aside the award,

or if he has refused such application,

they shall decide in accordance with the award of the majority of the arbitrators,

and shall fix the amount to be allowed for the expenses of the arbitration, and direct by and to whom, and in what manner the same shall be paid.

31. Such decision shall not be open to appeal, and shall be at once carried out;

Bar to appeal and suit in civil court.

and no civil court shall entertain any suit for the purpose of setting it aside or against the arbitrators on account of their award.

CHAPTER III.

Settlement of forest privileges for the convenience of the people.

32. Whenever the officers constituting the reserve deem it desirable to grant forest privileges in an ordinary reserve under Section 11, or in a special reserve, as an exceptional case, under Section 10, the officers aforesaid shall determine, define, and record every privilege so granted. The definition of privileges shall be in respect of the matters hereinafter provided for the definition of rights.

Conditions attached to privileges.

Unless otherwise specially directed by the officers constituting the reserve, and under sanction of Local Government, all forest privileges granted shall be for the personal or individual use of the grantee, and not for sale and merchandise.

All privileges shall be granted subject to the condition that their exercise shall cease in any portion of the forests closed for operations of reproduction, planting, or conservancy during the period of such closure.

CHAPTER IV.

Settlement of forest rights.

33. When the constitution of a reserve is proposed, a proclamation shall be issued in the vicinity, requiring all persons who claim to have forest rights in the forest to submit their claims to the officers charged with the constitution of the reserve within three months of the date of such proclamation.

Settlement of forest rights.

Provided that any person may submit such claim at any time within three years after the expiry of the three months, and on showing sufficient cause why it was not submitted before.

Limitation.

34. If on the submission of the claims the right is admitted by the officers constituting the reserve, such right shall be recorded, as it is admitted to exist.

Record of right if admitted.

If the right is not admitted by the officers aforesaid, the party claiming it shall be referred to establish his right in the civil court.

If not admitted.

35. If the right as admitted or decreed by the civil court is undefined, the officers charged with the constitution of the reserve shall define it.

Definition of rights.

Such definition shall consist in determining the following matters, according to the nature of the right to be defined :—

The quantity and description of timber, wood, brush-wood, bamboos, or other produce of forest, or its soil, to which the right-holder is entitled annually, or at certain seasons.

The time and place at which such produce shall be supplied, the manner in which it shall be obtained, and the extent and situation of the portions of forest which shall be assigned for the collection of such produce.

The extent and situation of the portions of forests which shall be assigned every season to those entitled thereto for pasturing of cattle, and the number of heads of cattle which may be sent into the forest to graze.

The date on which the season for pasturing cattle shall commence or close.

The portions of the forest in which hunting, shooting, or fishing may be practised, and the seasons during which such portions are open for the purpose.

36. Any person considering himself aggrieved by such definition may appeal to the chief revenue authority, who may direct further enquiry, hear the parties, and then either modify or confirm the definition, and his decision shall be final.

Remedy against definition.

37. Whenever it appears to the officers charged with the constitution of the reserve necessary for the preservation of the forests that any forest right shall be extinguished or commuted, they shall prepare a statement of the forest rights as they exist, and shall submit it to the Local Government, who may thereon order the extinction or commutation.

Extinction or commutation of forest rights.

Provided that in all cases where such commutation or extinction is ordered, full and fair compensation shall be given to communities or persons whose rights are affected.

Such compensation shall consist, either in the grant of land or of forest rights in other forests or Government lands, or in the payment of a sum of money, or in periodical payments, or in the abandonment of the right of Government in any forest or other land in favour of the person affected, or in the remission or reduction of any taxes or other Government demand, or in any of these conjointly.

Compensation in what to exist.

38. When such compensation has to be awarded, the officers charged with the constitution of the forest shall issue a notice to the right-holder or other person interested, requiring him to appear personally or by agent before the officer aforesaid at a time and place therein mentioned.

Procedure.

39. The provisions of Act X of 1870, or other law for the time being in force relating to the award of compensation, the reference to the court in case such award is not accepted, the appointment of assessors, &c., shall, *mutatis mutandis*, apply to all cases of extinction or commutation of right under this Act.

40. In any reserved forest, the officers aforesaid may, with the previous sanction of the Local Government, determine what roads and pathways shall be authorized for *public* traffic, and may cause all other roads and pathways to be closed either permanently or for a time only.

Roads for traffic in reserved forests.

Public notice of the closing of any existing road or pathway shall be duly given.

41. When the constitution of a reserve is complete, a report of the whole proceedings, together with the record of boundaries, the record of the settlement of rights, and adjustment of privileges, shall be submitted to the Local Government, who may confirm it or modify it in any particular: *Provided* that no right already settled shall be abridged by such modification.

Report of the constitution of the reserve.

42. The record, together with the orders of Government, shall be deposited in the office of Conservator of Forests, and shall have the same force of evidence as a settlement record.

Deposit of record.

43. From and after the date on which the constitution of a reserve has been approved by the Local Government, no fresh forest rights whatever shall accrue or be acquired in such forest.

No fresh rights to accrue.

44. Any forest of the 1st or 2nd class specified in Section 8, which before the Act came into force has been demarcated, may without further procedure be constituted as a reserve, provided that if no record of boundaries has been made such record shall be drawn up, and if no settlement of rights and privileges has been made, such settlement shall be made in the manner provided by Sections 32 to 41, inclusive.

Forests demarcated before this Act.

45. Any forest of the 3rd class, specified in Section 8, may be demarcated and its boundaries determined in the manner provided in the Act for forests of the 1st and 2nd classes ; and if the right of Government is such as to give it a complete power of management, such forest may be constituted as a reserve of either sort, provided that the right of the proprietor therein is not infringed.

Forests of the 3rd class.

In all other cases, unless otherwise arranged by consent of the parties, the right of Government shall be defined and recorded in the manner provided by Section 35 ; and subject to the maintenance of the Government rights, the forest shall be at the unrestricted disposal of the owner thereof.

46. Lists of all reserves, and of all forests of the 1st, 2nd, and 3rd classes demarcated under Sections 14, 15, and 45, shall be from time to time published in Gazette,

Publication of lists of reserves.

47. No reserve of either class shall be disforested or alienated either temporarily or permanently without the sanction in every instance of the Government of India.

Alienation of reserves.

48. All other forests not being reserves shall be open to be disforested, or given up to land cultivation to form village forests made over to village communities, or for other purposes by order of the Local Government; but until such order is issued, they shall continue subject to the provisions of Part 1 of the Act, and to such rules made under it as may be applicable thereto.

Alienation of other forests.

Forest lands made over to village communities shall be subject to such rules as may be applied to them in pursuance of Section 51.

Provided that no forest on any hill slope or at the source of any stream or other water, and no forest-producing trees in the plains, shall be disforested until it has been inspected by a Forest Officer, and it is certified by such officer that there is no objection to such disforesting.

49. Whenever it shall appear to the Local Government desirable on public grounds to expropriate any land for forest purposes, or any forest in which Government has or has not any forest rights, such expropriation shall be effected according to the provisions of the Land Acquisition Act, 1870, or other law for time being in force.

Acquisition of land for forest purposes.

50. Lands and forests so acquired may thereon be declared reserved forest of either class; the boundaries of such tracts shall be fixed, and, if necessary, a settlement of forest privileges made in the manner provided by Sections 14 to 32; both inclusive.

Treatment of such lands.

CHAPTER V.

Of forest rules to be made under this Act.

51. The Local Government may from time to time, with the sanction of the Government of India, make rules consistent with this Act for the protection and management of any forest, or any class or kind of forest, to which this Act applies, or of any groves, avenues, or trees along public roads, canals, around public wells, buildings, camping grounds, or other public places, or any groves or trees deemed sacred by any class of people, or growing around any tomb or sacred place.

Power to make rules.

Such rules may be made in respect of the following matters:—

First.—The protection of the forest—

By providing for the inspection and maintenance of boundary marks, and the keeping clear of boundary lines, the fencing of the forest, the protection of the forest against fire.

By prohibiting the marking, girdling, felling, and lopping, barking, tapping for gum, resin, wood-oil, or caoutchouc, or stripping off the leaves, or otherwise injuring any trees, shrubs, and plants.

By regulating or prohibiting the carrying or kindling of fire within the forest.

By prohibiting the collection and removal of leaves, fruits, grass, wax, resin, honey, tusks, horns, skins, and hides, stones, lime, dead leaves, and soil, or any other produce of the forest; by prohibiting the carrying of axes, saws, or other implements for cutting wood or timber except on roads authorized for public traffic, within the limits of the forest; by prohibiting the ingress into and passage through, such forests, except on roads authorized for public traffic; by prohibiting or regulating cultivation and the burning of lime and charcoal, and the pasturing of cattle within, or the passage of cattle through, such forests.

Second.—The management, utilization, and reproduction of forests—

By prescribing plans of forest treatment, utilization, and cultivation, and the closing of portions of the forest pursuant to such plans.

By prescribing the agency by which, and the conditions under which, timber, wood, trees, shrubs, bamboos, and other produce of the forest or its soil shall be cut, collected, removed, utilized, manufactured, or converted on the spot, or sold; by prescribing the agency by which, and the conditions under which, the leasing of pasture shall be effected, and the dues which shall be paid for the pasture of cattle.

By prescribing certain marks which are to be applied to all wood or timber or other forest produce extracted from Government forests, and prohibiting the use of such marks on any other timber or forest produce.

Third.—The protection of fish, animals, and birds within the forests—

By prohibiting the capture and destruction of fish during certain seasons of the year; by prohibiting the use, for fishing of certain descriptions, of nets, baskets, or other implements; by regulating or prohibiting the construction and use of weirs, dams, and other appliances for the capture of fish; and by prohibiting or restricting the use of drugs or other substances for the purpose of killing or intoxicating fish in rivers, ponds, and tanks; by prohibiting the killing certain kinds of animals, and the killing or taking the eggs of certain kinds of birds; by regulating or prohibiting the hunting, shooting, capture, or destruction of certain kinds of game during certain seasons of the year, and leasing or selling the right of hunting, shooting, and fishing; by regulating or prohibiting the capture or destruction of elephants or other animals at all times of the year.

Fourth.—The protection of groves, avenues, and trees, by prohibiting the felling, lopping, barking, stripping off leaves, tapping for gum, resin, or oil, or burning or otherwise injuring trees within a certain distance of any public road, river, well, or spring, or of any public tomb, building, sacred place, or camping ground.

Fifth.—The maintenance of forest roads and bridges,—by directing the levy of tolls from passengers or vehicles, &c., traversing such roads and bridges.

The Local Government may, with sanction as aforesaid, cancel, amend, or add to such rules. All rules made under this section, and all orders cancelling, amending or adding to the same, shall be published in the official *Gazette*, and shall have the force of law.

52. No rule made under this section shall affect or abridge any existing forest right
Saving of rights. (which has not already been extinguished or commuted under provision of this Act), nor shall it affect any forest privilege granted in accordance with Section 32, otherwise than in pursuance of the said section, as regards portions of the forest closed for treatment or reproduction.

PART II.

OF THE TRANSPORT OF TIMBER AND OTHER FOREST PRODUCE BY LAND OR WATER, AND THE LEVY OF DUTY AND ROYALTY THEREON.

CHAPTER VI.

Of duty on timber.

53. The Government of India may authorize the levying of duty on timber and other
Levy of duty. forest produce brought from any forest situated within British territory, or from any forest situated beyond the frontier of British territory, and may, by notification in the Official Gazette, prescribe the rates of such duty, the manner and place of levy, and the persons by whom it shall be levied.

Provided that if the duty is directed to be levied *ad valorem*, it shall not exceed 8 *per centum.*

Provided also that, if the duty is directed to be levied on each log of the said timber, it shall not exceed two rupees twelve annas per log of five feet in middle girth and upwards, and one rupee six annas per log of less than five feet in middle girth.

In every case in which such duty is directed to be levied *ad valorem*, the Government of India may from time to time by like notification fix the value on which the duty shall be assessed.

54. Government shall have a lien on all such timber or other produce for the payment of
Lien for duty. such duty; and in the event of such duty not being paid within a certain time, to be fixed by notification as aforesaid, the timber or such portion of it as may suffice may be sold to realize the duty (the balance realized, if any, over and above the duty being returned to the owner).

CHAPTER VII.

Of the protection of timber in transit, and of rules to be made therefor.

55. The control of all rivers and shores, as regards the floating of timber and of other
Control of rivers, &c. forest produce, and of such timber and other forest produce while in transit by land or by water, and of all Government timber depôts, is vested in the Conservator of Forests and in the Forest Officers of Government.

56. Except as provided in Section 66, no timber or other forest produce which, under rules hereinafter provided to be made, is subject to examination, or has to pay any duty or royalty, or requires to be marked before passing out of the control of the Forest Officer, or which is stranded, sunk, or adrift in any river, or on the sea-shore of any part of British territory, shall be marked, nor have any mark on it effaced or altered, nor shall it be converted, sawn, split, chipped, hollowed, cut into pieces, burnt, concealed, removed, sold, or otherwise disposed of, until it has been lawfully examined, paid for, passed, and marked, as the case may be, or until the special permission in writing of the Conservator of Forests or of the officer appointed to the charge of the forest division has been obtained.

57. Whenever it may appear necessary, the Local Government may fix, by notification in
Salvage services. the Official Gazette, the rates of reward to be paid for salvage services in respect of wood, timber, and bamboos; and no person who shall have saved any drift timber shall be entitled to keep the same, but shall, on tender of the sanctioned rate of reward, deliver up such timber to such Forest Officer or other person as may be authorized by the Conservator of Forests to receive the same.

58. All timber, wood, and bamboos found adrift on any river, or off the sea-coast of any part of British territory (to which this Act applies), and all unmarked wood and timber, or wood or timber on which the marks have been obliterated, altered, or defaced by fire or otherwise, and all timber beached, stranded or sunk in the rivers, or on the sea-shore as aforesaid,

shall be deemed to be the property of Government until or unless any person prove his right and title thereto.

Such timber may be collected by any Forest Officer or other person specially authorized in that behalf by the Conservator of Forests, and may be brought to such Government timber denôts as the Conservator of Forests may from time to time notify as depôts for drift timber. Public notice shall from time to time be given of such drift and unmarked wood and timber and bamboos so collected. Such notices shall not be issued in regard to timber, wood, and bamboos found adrift off the sea-const; but such timber, wood, and bamboos may be disposed of as the Local Government may from time to time direct. A period shall be fixed in the said notice of not less than three months within which claimants may prove their title to such wood, timber, or bamboos.

No wood or timber awarded to claimants shall be delivered until payment of any royalty or salvage which may be due, or other expenses which may have been incurred on account of such timber, shall have been made, and Government shall have a lien on such wood or timber for the payment of all such royalty, salvage, or other expenses incurred. On and after the expiry of the period fixed in the notice, all wood, timber, or bamboos to which no claim has been preferred, or to which no right or title has been established, shall become the property of Government, and may be sold or otherwise disposed of as the Local Government may from time to time direct.

Provided that nothing in this section shall be held to affect or abridge any *bonâ fide* right existing in any place, entitling any person or class of persons to appropriate to their own use any unmarked drift timber, wood, and bamboos generally, or any class or kind of such drift timber, wood, and bamboos: Provided also that the Local Government may, whenever it appears desirable, make a general order authorizing any person or class of persons to appropriate certain kinds of drift timber, wood, and bamboos.

59. All claims of ownership of timber, wood, and bamboos collected under the last section and brought to a Government timber depôt shall be decided by the Conservator of Forests or such Forest Officer as the Local Government may appoint in that behalf: Provided that such officer may, if he think fit, decline to decide the case himself, and refer the parties claiming to the court having jurisdiction in such cases. From all decisions of the Forest Officer, when the timber exceeds Rs. 500 in value, an appeal shall lie to the Commissioner of the division.

Claims to drift.

60. The Local Government may, with the sanction of the Government of India, from time to time appoint one or more officers to form a special court for the decision of claims to the ownership of timber or other forest produce, and may define the powers and limits of jurisdiction of such court, and lay down rules to guide the procedure of such court, and may determine the court or other authority to which an appeal shall lie, and the mode and time of presenting such appeal. Whenever such special court shall be appointed, the jurisdiction of the ordinary civil courts shall cease in respect of the matters provided for in this section within the jurisdiction of the said special court.

The Local Government may appoint special courts for the decision of claims to the ownership of timber and other forest produce.

61. The Local Government may from time to time, with the sanction of the Government of India, make rules consistent with this Act for the control of timber and other forest produce while in transit by land or water, whether such timber or forest produce comes from Government forests or otherwise.

Timber rules to be made.

Such rules may be made in respect of the following matters:—

The time and manner of floating timber, wood, and other forest produce. The prohibition of the closing up or obstructing of the channel or banks of any rivers which are used for the transport of timber or other forest produce, and the prohibition of throwing any kind of refuse into such rivers; prohibiting the stoppage of, or interference with, timber in transit by persons other than Forest Officers, or the levy of tolls or fees by such persons on such timber.

The use or carrying within certain limits of marking-hammers or tools whose sole or chief use is for making or altering or effacing marks on timber.

The registration of marks used as property-marks for timber or other forest produce, or implements for affixing such marks; the prohibition of registration in certain cases; the restriction of the number of marks to be registered for each person; the levy of fees for such registration and the time for which such registration shall hold good.

Prohibiting the use of certain marks used by Government by persons not being Forest Officers.

The collection of timber which from any cause is floating loose and without control; the issue or refusal of licenses for such collection; the collection of timber by Government Forest Officers, when such license is refused; the control of timber collectors and of the marking-hammers and boats used in collecting timber.

The salvage of timber and the control of salvors; the issue of licenses and marking-hammers to salvors; the treatment of timber salved without authority [*vide Rule 18 of the Burma Revised Code*].

The stoppage and examination of timber or other forest produce during transit by land or water; the issuing of passes for the same; the levy of fees for the issues of such passes; the form of such passes; the manner of levying royalty lawfully payable on timber and other forest produce; the detention of timber or other forest produce on which duty or royalty is payable, or which is not covered by a pass, or is suspected to be the subject of an offence against any law or rules for the time being in force; the reporting of timber and other forest produce; the production of certain passes or proof of ownership of timber or other forest produce; the places to which timber and forest produce has to be taken for the purpose of payment of duty or royalty; the registration of such timber or other forest produce at such places; the issue of certificates of registration and grant of extracts from the registers; the time and manner of removal of timber and other forest produce from such places, and the consequences of non-removal within a specified time; the registration of agents transacting business on behalf of timber owners.

The management and control of Government timber depôts; the control of persons employed therein, whether Government servants or not; the protection of public and private property in such depôts; and the responsibility of Government officers and others for its safe keeping.

Prohibiting the export from, or import to, certain stations, ports, or landing places of timber or other forest produce. Prescribing certain routes for the export and import of timber and forest produce, and prohibiting the use of certain routes for such export or import.

Requiring persons removing trees, brush-wood, and bamboos, and other forest produce generally, or of certain kinds only, from private forests, to obtain a pass or license from a Forest Officer before such trees or other forest produce are removed beyond the limits of the forest, and by regulating the issue of, and the conditions attached to, such licenses or passes.

Prescribing the duty of Forest Officers, as regards the prevention of offences against this Act, and the rules made pursuant to it; the reporting, laying complaints of, and prosecuting cases of such offences, and such other matters connected with the due execution of their duty in giving effect to the provisions of this Act and the rules made under it, as the Local Government may think fit to regulate.

The Local Government may from time to time, with sanction as aforesaid, cancel, alter, and add to such rules. All rules made under this section, and all orders cancelling, altering, or adding to the same, shall be published in the official Gazette, and shall have the force of law.

PART III.

OF THE PREVENTION AND PUNISHMENT OF OFFENCES AGAINST THIS ACT, &c.

CHAPTER VIII.

Of the prevention of offences and cattle-trespass.

62. It shall be the duty of every Police or Forest Officer to prevent, and he may interpose for the purpose of preventing, the commission of any offence punishable under the rules made in pursuance of this Act, or any breach of the provisions thereof.

Duties of Police and Forest Officers.

63. It shall be the duty of all communities or other persons who may have forest rights or privileges in any forest to which this Act applies, and of all persons who may have acquired by lease or otherwise the right of cutting, using, or removing timber, wood, and other forest produce, or of pasturing cattle in such forest, to assist to the utmost of their ability in preventing the occurrence of fires in the forest in which they have any right or interest as aforesaid. In the event of a fire occurring in such forest, it shall be the duty of all such communities or persons to aid in extinguishing the same, and to use due diligence in endeavouring to detect and bring to justice any person or persons who may be suspected of having been concerned in causing such fire.

Duties of persons who have forest rights.

It shall also be the duty of the communities or persons as aforesaid to prevent the commission of any offence punishable under the rules made under this Act, or any breach of the provisions thereof. And they shall use due diligence in endeavouring to detect and bring to justice any person or persons who may be suspected of having been concerned in the said offence.

64. Any Police or Forest Officer may seize and detain any cattle found unlawfully straying or doing damage to any forest to which this Act applies.

Cattle-trespass.

The provisions of the Act No. I of 1871 shall apply to such cattle-trespass, *provided* always that the Local Government may prescribe fines to be levied for cattle-trespass otherwise than as directed by Act I of 1871, and proportional to the damage which the different sorts of cattle do to the forest, and may prescribe that the fines be increased in the case of second offence, or if the trespass is between sun-set and sun-rise.

65. The Local Government may from time to time authorize Forest Officers to establish cattle-pounds in places where it may be deemed necessary; such pounds shall be subject to the provisions of the Act No. I of 1871, or other Act as aforesaid.

Establishment of cattle-pounds.

CHAPTER IX.

Penalties, &c.

Penalties. 66. The Local Government may prescribe penalties for offences against the provisions of this Act, or of any rule made in pursuance of it, or for any abetment of such an offence (within the meaning of the Indian Penal Code). Such penalty shall not exceed six months' imprisonment of either description, or Rs. 500 fine, or both.

The penalty may be prescribed in the form of an extra payment of certain fees, duty, or royalty, where such penalty is more convenient; and in default of payment of such penalty, imprisonment as aforesaid may be prescribed, or such confiscation as is provided by this Act.

The Local Government may prescribe a penalty not exceeding double the ordinary penalty, in cases where the offence is committed between sun-set and sun-rise, or after preparation for resistance to lawful authority, or if the offender has been previously convicted of the same offence.

Nothing in this Act shall be construed to prevent any person from being prosecuted under any other law for any act or omission which constitutes an offence against this Act, or the rules made under it, or from being liable under such other law to any higher punishment or penalty than that provided by the rules made under this Act: Provided that no person shall be punished twice for the same offence.

Power of arrest. 67. Any Forest or Police Officer may arrest without warrant any person committing an offence against this Act, or the rules made pursuant to it, under the following circumstances:—

(a).—If the offence is against Part II of this Act, or any rule made under Section 61 (timber transport, &c).

(b).—If the offence is of setting fire to the forest or wilful mischief prohibited by rules made under Section 51.

(c).—If the name and address of the offender cannot be ascertained, or if the offender gives a name and address which the officer arresting him reasonably suspects to be false.

In all other cases the arrest shall not be made without warrant: Provided that nothing in this section shall be held to exempt from arrest without warrant any person who would be liable to arrest under Section 92 of the Criminal Procedure Code.

Procedure after arrest. 68. Any person arrested without warrant on the ground that he has committed an offence against any provision of this Act, or rules made under it, shall without unnecessary delay be taken before a Magistrate who may, if he see reasonable cause, order such person to be detained in custody or held to bail until the case shall have been disposed of.

Jurisdiction. 69. Any Magistrate having local jurisdiction may upon complaint of the infringement of any provision of this Act, or of any rule made under it, or without complaint, if such infringement shall come to his knowledge, take cognizance of the same, and shall proceed as prescribed by the Code of Criminal Procedure for the investigation and trial of summons cases. Offences against this Act or the rules made in pursuance thereof may be tried summarily under the provisions of Section 222, Criminal Procedure Code.

Property connected with case under trial. 70. When any person is convicted of any offence against the provisions of this Act, or of any rule made under it, it shall be lawful for the court, when the trial is concluded, to make such order as appears right for the disposal of any timber, wood, forest produce, or other property regarding which an offence appears to have been committed, notwithstanding that such property cannot be, from its nature, produced in court before it.

In any case in which any offence which would be punishable under this Act is tried under the Penal Code or other law, the provisions of Section 418, Criminal Procedure Code, shall apply to the disposal of property connected with the case, notwithstanding that such property is not, owing to its bulk or otherwise, produced before the court.

Confiscation. 71. All timber or other forest produce in respect of which any infringement of this Act, or of the rules made under it, has been, or is suspected to have been, committed, or other produce, or on which any mark has been obliterated or altered, or which has been wrongfully marked, or for which the duty or royalty lawfully payable thereon has been evaded or attempted to be evaded, or which is not covered by a pass or other proof of ownership as prescribed by the rules made under this Act; and all tools, boats, carts, and cattle used in infringing any of the provisions of this Act, or of the rules made under it, shall be liable to confiscation in addition to any penalty to which the owner or person in possession may be liable.

Such confiscation shall not be awarded in addition to any other penalties in cases where the payment of an extra duty or royalty by way of penalty is imposed in pursuance of Section 66 of this Act.

Procedure in case of property liable to confiscation. 72. When any timber or other property is liable to confiscation under the last section, it may be seized by any Police or Forest Officer. On such seizure being made, a distinctive mark, indicating the same, shall be put on the property seized, and the officer shall, as soon as may be, apply for the confiscation of the property to any Magistrate having local jurisdiction at the place of seizure.

73. The Magistrate shall then summon the owner, or the person found in possession of
Procedure in confiscation cases. such property or person suspected to have been concerned with the same, and upon his appearance, or, in
default thereof, in his absence, shall examine into the cause of seizure of such property, and
may adjudge the confiscation thereof; such property shall thereupon belong to and vest in
Her Majesty, and the court shall issue a warrant to that effect, directing the Forest Officer
having local charge, to hold the property confiscated at the disposal of the Local Government.

Appeal in criminal cases. 74. An appeal shall lie from every order of confiscation where the property exceeds Rs. 500 in value, to the
Commissioner of the division.

75. A portion, not exceeding one-half of any fine or money penalty awarded under this
Reward to informer. Act or other law, on a trial for an offence which is
punishable under this Act, may be awarded by order of
the court to any person, not being a Forest Officer, on whose information the offenders were
brought to justice.

76. The provisions of the Code of Criminal Procedure shall apply to the trial of all cases
Procedure in cases under this Act. and appeals under this Act, and to the admission of
appeals in cases where an appeal is not specially provided
for in this Act.

77. It shall be lawful for the Magistrate of a district or division of a district, on applica-
Search for stolen and contraband produce. tion of a Forest or Police Officer stating his belief that
stolen timber or other forest produce, or that timber or
other forest produce liable to duty or royalty, or prohibited from export or import, is secreted
in any place in such district or division, to issue a warrant to search for such goods. Such
warrant shall be executed in the same way and shall have the same effect as a search warrant
issued under the Code of Criminal Procedure.

78. The Local Government may determine on whom the duty of maintaining forest
Penalty for erasing boundary-marks. boundary-marks lies—whether on Government or other
person, according to the nature of the forest. Any
person wilfully erasing, removing, or damaging any boundary-mark, may, on conviction before
a Magistrate, be ordered to pay such sum, not exceeding Rs. 50 for each mark so erased,
removed, or damaged, as may be sufficient to defray the expense of restoring the boundary-
mark, and of rewarding the informer through whom the conviction was obtained.

"STATEMENT OF OBJECTS AND REASONS," OR EXPLANATORY MEMO- RANDUM, BY B. H. BADEN-POWELL, OFFICIATING INSPECTOR GENE- RAL OF FORESTS: TO ACCOMPANY HIS PROPOSED DRAFT FOR A FOREST LAW.

It would be tedious to go over the defects of Act VII of 1865. Those defects are admitted; they are also detailed—at any rate the chief of them—in a paper printed in the report of
the Forest Conference of 1873-74.

As far back as 1868 a revised Bill was put forward, and was circulated for opinions to
Local Governments. On receipt of all the replies, a further revised draft was prepared, which
was again circulated in 1871; and at last, after long delay, the opinions on it have been
received.

A careful study of those opinions has been my chief stand-by in attempting to frame a
Bill.

I have also carefully considered all the drafts of rules that exist, to see that as many of
their provisions as are likely to stand are covered by the terms of the Bill. But in point of
fact, the rules in force are all of them indifferently drafted. The Burma Code is condemned;
the Oudh Code is notoriously imperfect; so is that of Bengal. The North-Western Provinces
have a Code just drafted, and not issued; but this is, it is hoped, tolerably efficient, and will at
any rate be legal, if a law is passed as proposed. In the Punjab no general Code has been
issued; a few rules for river management—very imperfect—were promulgated under Act VII.
The rules for Mysore are not good; those of Coorg are brief, and believed to be imperfect.
The Central Provinces rules are badly worded; but as much of them as would be likely to be
retained would be legal under the proposed law.

I think I can safely answer for the present Bill—

(1.)—That it meets all the really valid objections to the latest circulated draft.
(2.)—That it covers the revised Code of Rules for Burma and the North-Western
Provinces, and will be found to meet every rule that will be framed in other
provinces.
(3.)—It is based on principles that were discussed and approved at the conference of
1873-74.

The Bill starts with a series of preliminary sections. It will be observed that it is designed to repeal all the old laws, and have one Code complete in itself. The Burma Act XIII
of 1873 will disappear, except so much of it as indemnifies for past Acts.

It is proposed to repeal all rules; this Mr. Brandis's draft did not: but I am convinced
that it is much easier to do so. There is not a single Code that I am aware of that is either
properly drafted or satisfactory; it is much better, therefore, that old material should be swept

away. Besides, if any existing rules are good, what is easier than to republish them uniformly under the new law? Then, as each province issues its own Code, copies can be printed and circulated, and volumes compiled. There is, however, no object in inserting in the Bill any provision requiring the collection and publication of forest rules; that can be done by the authority of Government, or even by any private author.

Section 4 defines the term " forest." I think this is the best definition, because it is the most natural. Practically it amounts to this, that everything is called forest which is naturally and obviously understood to be forest or "jungle"; but that if Government wishes to call and treat as forest, a plot of bare land taken up to form a plantation, and so forth, then such plot shall be specially notified as forest, and become subject to forest *régimé*, pursuant to the notification. I believe nothing will be found simpler than this.

The allusion to the "property-mark" is based on a suggestion in the Bombay Bill.

The question also arose in the recent Burma permit frauds. Here a number of standing trees were fraudulently marked with the Government hammer; but as standing trees, they were not *moveable property*, therefore the Penal Code did not apply.

In the definition clause, too, the all-important distinction between forest *privileges* and forest rights has been maintained; the omission of this subject gave rise to no little confusion in commenting on Mr. Brandis's draft. For instance, in Section 27, we *meant* that forest *privileges* should not be alienable; but as worded, the section refers to *rights*, which may or may not be alienable according to their inherent nature.

Rights are strict legal rights, which unquestionably exist, and in some instances have been expressly recorded, in land settlement records. These Government cannot interfere with, unless in cases of urgent necessity it compensates and expropriates under the provisions of the "Acquisition Act," X of 1870.

Privileges are concessions of the use of grazing, firewood, small wood, &c., which, though not claimable as of legal right, are always granted by the policy of the Government for the convenience of the people. These, however, have been *always* regulated in some respects, and it will be found that either the existing practice or the existing rules of every province recognizes the following limitations:—

(1.)—Privileges may be stopped in portions of the forest which Government wishes to close for planting or reproduction.

(2.)—They are always given for the convenience of individuals, villages, &c., but are not intended to be sold and bartered. People are allowed to take firewood— *e. g.*, for their own house—but not to cut cart-loads and sell to a bazar dealer or a railway depôt.

This is only reasonable. I may here explain (for the benefit of Berar Forest Officers) that this does not militate against the practice of the Koorkoos. These people are allowed to take produce out of the forest and sell it; but a *Government royalty is levied on* all they offer for sale at certain choukis. This is not a case therefore of " privilege."

Section 7.—I print this between brackets. I doubt the advisability of having the clauses ·in the Act; I think such provisions had better form portions of a Code of " service regulations," which are in fact terms of the service contract between the State and·its servants, and should not be put into an Act.

There exists many rules about land-owning, holding shares, &c., relating to different officials and services, but I am not aware that any of them find a place in a Code enacted by the Legislature.

These preliminaries being disposed of, I will proceed with the Bill.

It is divided into three parts or main divisions:—

    I.—Relates purely to forests.

    II.—Relates purely to the import and export and transit of forest produce, and the levy of duty, &c.

    III.—Relates to penalties and prevention of offences.

PART I.

Section 8.—All forests in India come under one of these three classes. The first is the commonest, because, observe, we are talking of *rights*, not of what Government *will out of kindness allow.*

The second is rare, and is, I believe, confined to the Punjab,—notably in·Kulú, where forest rights are entered on the records.

The third class also occurs in the Punjab, in Kangra; they are the result of the action of bygone years, when officers thought it was enough to assert the right of the State to the *growing trees* of all or of certain kinds.

I believe this class is very common also in Bombay. Blackwood and teak are Government property when they grow on certain lands which are otherwise not Government property. The forest is in fact a private estate with Government rights in it. Kangra in the Punjab is exactly. the same.

The Central Provinces Government took an objection to the former draft,·which I cannot think a very sound one, that because they had only forests of the first class, and not of the second and third, therefore they wanted a separate Act; and one general one, which possibly contained provisions of no use to them, was called cumbersome, &c.

I may shortly dispose of this by observing, that as it is the first class of forests and the settlement of *privileges* (which are extensively allowed in the Central Provinces) that give the chief trouble, it will not be found that this Bill contains more than two or three sections, at the outside, which will be superfluous in the Central Provinces. The advantages of one general law and policy of forest administration are *so* great that I think no more need be said on the subject.

Section 9.—Any forests of the first two classes is Government property, soil and all; so that, provided existing rights are respected inside them, and privileges duly provided for where needed, there can be no objection to constitute them "reserved" forests.

The third class, on the other hand, presents difficulties, and has to be specially provided for: of this presently.

At present forests are called "reserved" and "unreserved" or "open," but these names are indifferent in themselves, have different meanings in different provinces (*vide* General Administration Report for 1872-73, Chapter II), and are liable to be misunderstood, because if we say "unreserved," it implies that no kind of protection is extended; whereas to the "unreserved forest" in the North-Western Provinces, Punjab, and Berar, protection *is* extended; and in the Central Provinces, where it has not as yet been so, the condition of the forests is such as to be exciting alarm (*vide* Central Provinces Forest Report, 1872-73).

I propose, therefore, to use the terms "special reserve" and "ordinary reserve"—the one implying a close tract (just as at present) in which privileges are not allowed, if possible, to help it; and "ordinary reserve" ("unreserve" of Berar, &c.), implying a tract of forest to be *kept as forest*,—not to be destroyed, but chiefly so for the benefit of the people (where the people have not already got areas of common or waste land made over absolutely to them).

All forest area will be organized, then, as either a "special reserve" (close forest) or of "ordinary reserve" (open forest). Such forest as is not made into either, will be left subject to the general protection of the law, as afterwards stated. The constitution of special and ordinary reserve is in fact, in principle, exactly what is done now, only that it avoids the objection attaching to the "unreserve," as it exists in some places.

How much land *ought* to be specially reserved, and how much generally, depends on a study of the effect of forests on the soil and climate, and on a knowledge of how much material, and of what sort, the State desires to produce for home consumption and for export; and what area—of what class of forest, under what treatment—will permanently (and without tranching on capital) produce that quantity and quality.

It is the duty of the Forest Department—and one which I hope we shall not be long before we very earnestly undertake—to study that; the law cannot prescribe it. Meanwhile, assuming that certain forest lands are available, and may be placed under treatment, how is the forest to be constituted as such? Obviously three things have to be done:

    (1)—To fix the boundaries.
    (2)—To provide for the convenience of the people (*i. e.*, what *privileges* are required).
    (3)—To consider not only the general question of convenience, but the actual legal *rights* existing in the forest.

As regards the first of these, *very* often the boundary of a forest is a mere matter of convenience. Out of a vast area of jungle a line is drawn to take in the best parts; in that case no private person or community has anything to say to it; it is a matter of *convenience*; but it often happens also that the boundary of the forest adjoins a cultivated estate, a private forest, or a tract which is the subject of some private *right*.

Then, the boundary may be *disputed*; the Government may claim one line, the adjoining proprietor may insist on another. Mr. Brandis's draft proposed a "commission" to settle such questions.

All Local Governments nearly objected to this, and I am now convinced myself that it would not work.

The Bill therefore proposes a scheme, of which I claim two advantages:
    (1) It is suggested in *several* of the "opinions."
    (2) It has already worked admirably in two districts in the Punjab, and practically in Berar also.

I refer to the appointment of two experienced officers of Government. The Local Government has its whole staff to select from —one a Civil or Settlement Officer, the other a Forest Officer; and these two shall go out to the place, and shall effect what for brevity I may call a "forest settlement."

You cannot have a Civil Officer alone, for he does not know what the forest requirements are, and the result will be as in Bombay, where certain forest demarcations were made by Revenue Officers alone, that now prove utterly useless for any forest purpose whatever.

You cannot have a Forest Officer alone, because his more exclusive view of the subject, and the greater prominence which forest requirements naturally have in his eyes, might lead him to overlook the contentment of the neighbours. Therefore, take these both, and entrust them (Section 12) with the duty of what I have provisionally called "constituting or settling the reserve" as regards the three subjects already alluded to.

It may be asked, how will it be determined whether the reserve is to be a "special" or an "ordinary" one? I answer that no line can be drawn on paper; but on the ground it will be easy. Where the forest is good, or when it may be made good, if not so now,—where there is no great fear of oppressing the people—circumstances will indicate to the officers in charge to make it a special reserve. Where privileges *must* be allowed, or rights are numerous, then make it an "ordinary" reserve.

I know this is not a strictly scientific forest proposal; but we have to make a compromise, and try and meet both views as far as possible.

Section 13.—If, in determining on these and other questions, the two officers differ, then the chief revenue authority will decide, and there is to be an end of it.

In practice, with good and experienced officers, such differences will very rarely, if ever, arise.

CHAPTER II.

Section 14 needs no comment, nor *Section* 15; the forest may not in all cases be worth demarcating very expensively, nor indeed may the sort of treatment it is put under demand very careful demarcation at every point. It must be left to the officers actually on the spot to recommend *what extent of perfection* in demarcation is needed : that is the object.

Section 16.—But then of course, as already intimated, directly it is not a mere question of picking out of reserve out of the centre of a whole jungle, but we come in contact with cultivation, &c.; then boundary disputes between the forest interest and the private proprietor may arise. What then?

Why, as already explained, instead of the commission of Mr. Brandis's draft, I adopt *verbatim* the provisions of Act XIX of 1873, the North-Western Provinces Land Revenue Act, which embodies the provisions of Act I of 1847, &c., and empowers the Civil Settlement Officer to refer the matter to arbitration. The entire set of Sections 16 to 31, both inclusive, are copied from the Act of 1873, and give all the details about arbitration. It is the insertion of the details about arbitration here that make the present draft so long; if these 15 sections are excepted, it will be observed that the whole of the rest of the Code is of only 63 sections.

I presume there is no need for comment on provisions which are details already passed by the Legislature.

CHAPTER III.

Section 32.—Boundaries being laid down, the next question is, how is the convenience of the people provided for? Again, the officers charged with the duty make local inquiry, and determine (subject to review by the Local Government—Sections 12 and 41) what is necessary, only that they must record and define. This improvement on the old procedure is *absolutely a sine quâ non*.

The history of all forest management has shown that whenever you either deal with actual legal rights or with privileges which you concede, and which (subject to the *conditions* which you attach to them) are exercised like rights, you must define them, or they grow and extend to such a degree as utterly to swamp your forest. You begin by granting an undefined privilege to a hamlet of four houses to take wood for repairs of the houses; this right may be worth Rs. 5 a year, but being undefined, in time the four houses become 400, and the wood for repairs is no longer the mere poles that thinnings would supply without loss, but the whole produce of the forest is swallowed up, and the right is now worth Rs. 50,000! This is of course an extreme case, but shows what I mean. It is not enough to say "zemindars have the privilege of firewood." What zemindars? and how much wood? (the quantity may accumulate, but not be increased). It is not enough that such a village has grazing: how many houses, and how many head of cattle of each kind? Neglect of this will not only gradually render forest management impossible, but is inflicting a gross injury on those who are to come after us: and it is the duty of the State to act "from the stand-point of its longevity," and not work purely for the temporary convenience of to-day.

I have to apologize for dwelling on these common topics, to be found in treatises on forest political economy; but the subject has been overlooked hitherto. It is easy to define when you are making the "settlement," and very hard to get an opportunity afterwards.

CHAPTER IV.

Section 33.—Forest rights being matters beyond the option and judgment of the Settlement Officer, it is necessary to know what they are, so that a proclamation has to be issued for claimants to come forward.

As to the time to be allowed, after long consideration I think three months is *practically* ample. In the first place, in practice the rights are never exercised, except by people who are close at hand, and can easily attend when notice is given that a forest settlement will take place. If we do not have that, either the settlement work will be indefinitely prolonged or never finished at all. It may happen that a right-holder is away from home; and if so, save his right by allowing the case to be put in afterwards and specially dealt with.

Many persons so coming forward will no doubt, in a great number of instances, state as rights what are not really rights at all. Then the Settlement Officer will exercise a wise discretion and may find it fair to allow as *privilege* what he does not admit as a right.

If he admits the right, he will record it *as it exists* (Section 24).

If not, and if the claimant insists on the right, and declines to take it (supposing such is offered) as a privilege, then he must get a decree against Government in the ordinary civil court (Section 34).

Section 35.—In all cases the right *must* be defined. This is, as before said, the *sine quâ non* of effective forest management. Perhaps the court should be required by law to define the right at time of passing the decree; that is a detail which I may leave to consideration, but definition there must be.

What that definition is might be almost naturally understood; but so important is the subject, that it is, I think, better to specify.

Section 36.—If any one is aggrieved by this definition, let the chief revenue authority decide finally what the proper definition should be. This appeal will rarely be needed, because, as a rule, the right can be ascertained; there *cannot be such a thing as a right to which limits cannot be assigned.* Still it is possible that the Settlement Officers may define the number of cattle and the area of forest, &c., too closely, and an appeal may be necessary.

I do not see the use of retaining any separate power of "*regulating*" a right as Mr. Brandis's draft does; if the right is once accurately defined, we must put up with it, or else, if it is so bad that conservancy is impossible in the face of it, we must commute or ex-tinguish it.

"Regulation," it may be replied, is really a sort of restriction or half extinction of the right; in that case I say the provision about extinction applies, and compensation must be given.

Section 37.—When the right is so obnoxious as to threaten destruction to the forest, this has to be represented to the Government, and then Government, if it considers the case one of sufficient urgency, will act under Sections 37 and 38.

Section 39.—Agreeably to the uniform expression of opinion and to the views of the conference, I have proposed simply to apply the existing law, Act X of 1870, to the case, *mutatis mutandis.*

Section 40.—This section is necessary, as it will be observed that some forests have all sorts of tracks, not regular roads, running through them, and it is necessary to get the traffic into a proper line. This section deals only with roads that are used by the *public.* A *private* right of way is a forest right (definition clause), and if it has to be stopped, then the compensation section applies.

In closing *public* roads, the object is simply to confine traffic to the proper and necessary course, and not leave it to carts and cattle to strike out tracks of their own, as they do in Burma, wherever they please. Of course, no one would think of closing a "pucka" road or a regular village road; if they did, the Government would refuse to sanction it.

Section 41.—The settlement being now complete, Government reviews it and records its general assent or sanction; it may modify where necessary, but that modification cannot of course relate to matters that have been determined and recorded by consent of the parties, or have been decided by regular legal procedure.

Section 42.—Needs no remark.

Section 43.—This is obviously fair. Forest rights grow up by prescription, and as they are always rights by which the *public* property or benefit is diminished for the good of an in-dividual, while we are bound by equity to respect the present, we are equally bound not to bur-den posterity with new rights created in the same way. Of course, without this section it would be possible to prevent the growth of rights by continual action declaratory of the right of Government; but it is far simpler to put it down once for all as proposed.

Some of the opinions on Mr. Brandis's draft (which also contains this provision) misun-derstood it, and spoke about existing incumbents and the right dying with the holder. This is not intended; if the *right existing* is in its nature heritable, it will not be extinguished on the death of the holder by this section. Supposing, however, in a valuable forest 40 houses. (as already instanced) *now have* a right to wood worth in all Rs. 50 a year, it is forbidden to wrong those who shall come after us by allowing the 40 to swell to 400, and the burden on the public property to rise from Rs. 50 to (say) Rs. 5,000, or perhaps to swallow up the entire pro-ceeds of the forest.

It is remarked that we have to benefit by the experience of other countries; this result *has actually occurred* in more than one instance on the continent.

To people, therefore, who object to our guarding the future by such a provision, history teaches in vain.

Section 44.—Provides for cases in which demarcation of reserves has already been completed before the Act comes into force.

Section 45.—Deals with the case of third class forests.

Here the *soil* is the zemindar's; the trees on it are Government. It is quite right to -demarcate it, because of course the limit of the tree growth is determined thereby; but it would not be right to take the whole place in charge as if it were Government property. All we can do in such cases is to see that our trees are not injured, and protect the State right by such rules as effect the object without infringing other rights.

Practically, however, we always endeavour to put a stop to so awkward a state of things, by getting the people to consent to give up a portion of the forest absolutely to Government; the Government in return giving up its right to trees, and perhaps some cash, &c., into the bargain; then, by consent, the Government acquires the portion agreed on as a special reserve of Chapter I.

Section 46.—Is a detail perhaps hardly of much importance.

Section 47.—Is essential. The Government of India, as directing the forest policy at large, must have the sole right to permit the reduction of the constituted forest area.

Section 48.—Provides for the clearing and cultivation of all other Government forest land which, however, as long as it is not broken up, is under the general forest law.

It aims at introducing some sort of order into the procedure. At present, the proposals for giving up land are sent in, and there is no satisfactory method of deciding whether the land is to be given up or not. The clearings in Coorg, in Darjeeling, &c., are open to this remark especially.

This section will at least secure, that before, by a few strokes of the pen on the back of an "arzee," valuable forest, perhaps on a south face, or a steep slope, on which its retention is essential, is granted away, a proper examination be made.

Considering the ease with which forest is cut down, and the difficulty with which it is got to grow up again, something more should be done than merely regulating prices and terms of waste land sales.

I may here take occasion to remark that the provision of Mr. Brandis's draft, Section 13, will not be necessary under the law proposed. If a forest reserved by Government is treated entirely for the benefit of the village or individual, then the settlement, by recording the privileges to be exercised, will secure this. If, on the other hand, it is designed to give up a bit of forest absolutely to a village, &c., it can be done under Section 48, and such general conditions applied to the treatment of the forest so made over as the rules prescribe (*Confer.* Proposed Burma Forest Rules, at page 60 of my report on the Burma forest system, Rules 22 to 26: only of course this Chapter should be headed " of forests made over to village communities").

Section 49—Provides for the acquisition of land for plantations or forest purposes, and would of course include the case of a forest of the third class or a private forest, the treatment of which by strict rule was desirable on public grounds.

In Mr. Brandis's draft (Section 42) a separate case was made of forests which were to be preserved on account of their value in preserving the soil on slopes, &c.

But at the conference we agreed that it was better not to draw such distinction, but that whenever on any ground the Government desired to take up a private forest or a plot of ground, &c., it should do so only under the general law of " land acquisition"; hence there is no object in specifying all the details of Mr. Brandis's draft, Section 42.

CHAPTER V.

Section 51—Power to make rules. I think everything has been inserted.

The latter subject of the " second" clause is required by Bombay, and probably elsewhere.

The " fourth" clause is a novelty; but it is essential, I think, to protect these groves and tree avenues, which are now being made on lands which are not (like canals) under any special law of their own.

The " fifth" clause is also new. It is a suggestion made at the conference. In many places bridges and roads have been made by the Forest Department, and these have been at once thronged with public traffic; it seems right that Government should, if it chooses, put on a toll, so as to aid the forest funds in maintaining the road, &c.

Section 52—Needs no remark.

PART II.

This deals wholly with transit of timber by land or by water.

Section 53—Repeats the provisions of Act XIII of 1873, applicable to Burma, but which may be required in other provinces also.

Section 54—Provides the lien for duty, so that should timber on which duty has not been settled be sold in execution of a decree, the duty is a first charge; the subject principally concerns Burma.

CHAPTER VII.

Section 55—Needs no remark.

Section 56—This gives the key-stone of river management. Having once put the timber out of the forest into the water, until such time as it reaches depôt, it is taken under the guardianship of the State for several purposes:

(1).—To protect it from all sorts of fraud, timber theft, &c. .

(2).—To see that it is not material which has been obtained by the commission of an offence in the forest (thus indirectly serving forest protection also).

(3).—For revenue purposes, to collect duty, royalty, &c.

There is no possible object in allowing timber to be marked as a rule; that should be done before the timber is launched; if a special case occurs, why then the owner can get a special permission. But once recognize the right of *any* one to mark timber while in the river, and all prevention of crime is at an end. A wide door is thrown open to fraud. Every one shelters himself under the plea that he was marking *his own* timber.

Timber also cannot be "converted," so as to become unidentifiable, till it has been examined and passed, &c.

Section 57—Regulates salvage or the stoppage of timber to prevent its being carried away and lost by persons not being the owners.

Section 58—Provides the procedure regarding drift timber which is already in force, either legally or illegally, in most provinces.

Government has a *primâ facie* right to all timber that is ownerless or abandoned, and an orderly procedure in collecting this is required. In order to give owners a chance of identifying their timber and proving that it is not ownerless, &c., a regular notice is issued and procedure laid down for the settlement of claims.

The proviso at the end of the Section 58 is put in to meet the case of the Teesta and Brahmaputra rivers, where the people have the right to drift *pieces*, but probably not to timber. In other places also it may be desirable to allow people to appropriate certain kinds of waif pieces, &c.

Section 59—Is on the same subject.

Section 60—Is put in because it may be desirable in Burma to constitute such courts. A good deal of correspondence has already taken place. The section being permissive only, there can be no harm in the insertion of such a power.

Section 61—Empowers the making rules on all details about timber transport.

In preparing the section I have chiefly considered the requirements of Burma. In this province there is an enormous timber trade and a complete system of river management, both as regards timber from British forests and timber from beyond the frontier. There are regular stations at which the processes of duty-paying, examination, &c., are carried out; it is therefore quite certain that provisions which will cover the new Burma River Code, as approved by the Local Government, will cover every kind of provision which any river can require in India.

As regards the first clause, "prohibiting the stoppage," &c., is aimed at the evil described in the Bengal Government letter and enclosures.

The next clause about marking-hammers is required chiefly in Burma; also the clause about registration of marks.

The "collection" of timber, *i. e.*, by its owner, it is found convenient to distinguish from *salving*, which is effected usually by people *not* the owners.

The sixth clause relates not only to water transit, but also to land.

The seventh clause explains itself, and is chiefly aimed at the great revenue station of Kadoe in Burma.

The eighth clause is required in the North-Western Provinces, where the kham tehsil system is in force. These people get forest produce of certain kind, and bring it out by certain routes to toll-houses, where it is examined, pass issued, and the prescribed seigniorage collected. Often it happens that the forest is so situated that only certain routes are practicable,—then the system is complete; but it may happen that there are other routes available by which the toll-house might be evaded.

This applies also to all forests, such as Central Provinces and Berar, where the "naka" or toll-house system is in force.

In Bombay also, the provision about the produce of private forests is specially required. Details may be learnt from the correspondence with Bombay; but where the forests are interlaced with private holdings, Government gets robbed in all directions, and the produce cannot be recovered, because there is the plea ready that it came from a private forest.

These provisions apply to *all* timber, British or foreign—from private forests or Government.

It is obvious that private and foreign produce need, while in territory to which our law applies, to be protected and also to be examined, or very soon extensive smuggling would be perpetrated under colour of the material "coming from beyond the frontier," &c.

PART III.

PENALTIES, &C.

CHAPTER VIII.

Section 62—Is the same as the old law.

Section 63—Is adopted from Mr. Brandis's draft. It is but fair that persons who are *in* the forest,—in many cases allowed these of favour and kindness, and for their own benefit—and who, therefore, have exceptional facility for assisting Government, should be made to help when danger threatens.

Section 64—Applies to the Cattle Trespass Act; but for all *forest* purposes the fees fixed by Act I are so absurd that a special scale is necessary, which *proportions the fine* to the damage done. For instance, a cow does only a little damage in a plantation or young forest; a buffalo does much more, owing to his greater weight; a goat is the most destructive animal possible; yet, by the existing scale, trespass by a cow is heavily fined, and a goat has the most trifling fee; it ought to be just the reverse. So I would, propose to remedy this defect. Again, the present scale of fees is so light that in some valuable plantations at Lahore it has been found that the people drove in their cattle and let them be pounded, saying that it was cheaper to feed them like that and pay the fee!

Trespass repeated like this, or done during the night, should be more heavily punished.

CHAPTER IX.

Section 66.—The penalties are so arranged that where convenient it may be lawful to provide, *e. g.*, as in the Burma Code, "in case of non-report or incorrect reporting of timber at Kadoe, it shall be liable to four times the duty, &c."

For the rest, there is no reason why the penalty for breach of forest and timber rules should only be met with *fine*. It has been found in Burma that while timber thieves will pay a fine, the fear of jail, even for a few days, is a more powerful deterrent. It is therefore much wiser, considering that the penalty is first of all fixed by the Local Government for each offence separately, if it pleases, and then can be proportioned (within the limit of the rule) by the Magistrate; it seems much wiser to leave the Local Government the option of imposing fine or imprisonment, or both. Grave offences, such as wholesale mischief, extensive frauds, &c., should be prosecuted under the Penal Code, and not under the rules.

Section 67.—The power of arrest without warrant in all cases has stood on the law (Act VII of 1865) for nine years without any complaint of hardship; but as it *might* give rise to difficulty, it is better, as noted by the conference, to restrict such power—

(1) To cases under the *timber* rules, when *without such power no offender would ever be brought up at all.*

(2) To cases of wilful damage to the forest.

(3) To cases where the offender's name and address cannot be ascertained.

In all timber cases the fact that a warrant is applied for always leaks out somehow, and at once all traces of the offence disappear! (*Vide* Conference Report, page 19.)

Sections 68 and 69—Do not require comment.

Section 70—Is designed to remedy one of the great defects of Act VII, which Mr. Brandis's draft did not provide for.

Confiscation under Act VII would be applied both to cases of *real* confiscation, depriving an *owner* of his timber or his area, &c., and to those cases when it is simply taking property to which the accused had no title, and giving it back to the owner,—restoring stolen property in fact.

In the latter case there is no *confiscation* at all, and therefore Section 70 provides that when the trial is over .the property is to be restored to its owner, just as in Section 418, Criminal Procedure Code. And it is also provided that if in a trial under the Procedure Code the stolen property is not *produced*, nevertheless the rule in Section 418 applies. This actually came up in a trial at Tounghoo, in which one Mahomed Ali was accused of mischief and misappropriation in respect of about 800 logs of teak timber; he was convicted, and application was made for an order to restore the timber, the subject of the offence, under Section 418, Criminal Procedure Code. The objection that it was not produced before the court was raised by the Judge, but was not pressed by the defence; so the order was made.

Section 71.—This provides for the true or real confiscation. In the Punjab it is thought that confiscation need not be *additional* to the fine and imprisonment.

Of course, as long as the old *confused* confiscation rule applied, it was essential that the so-called confiscation should be additional, or else, if the Government recovered *its own tree* that was improperly felled, it could not punish the offender.

Now, however, that confiscation in the section means only real confiscation, that is, depriving an owner of his property, I do not see much need for its being additional; I have however, left it so, as the Local Government can always regulate the application of the double penalty and prevent its being worked harshly.

The only one case in which "confiscation" is applied to a case where the person in possession is not the owner, is when a quantity of timber is seized as having come out of the forest unlawfully, *e. g.*, timber is found in possession of the agent of a firm which holds permits to work certain forests. Clearly, by the state of the timber, it has been obtained contrary to the terms of the forest rules, and is not such as the permit gives a right to. Here, strictly speaking, the agent is not the owner, at the same time he is not personally accused of a breach of forest law, and therefore the timber could not be taken from him, as by Section 418, Criminal Procedure Code, in such a case the timber is taken under this confiscation section.

Section. 72—Regulates the procedure in case the officer of Government deems right to proceed for confiscation. In the recent great confiscation cases in Burma, the question constantly arose—when was the seizure made? what was the process of seizing? &c. Section 72 lays this down.

Section 73—Requires no comment.

Section 74.—Distinctly clears up the question of appeal. At the present it is not at all free from doubt whether there is any appeal for a confiscation case, because it is doubtful whether the Criminal Procedure Code deals with cases that directly resemble these confiscation cases.

Section 75.—Supplies an omission which is needed.

Section 76.—Requires no remark.

Section 77.—Is new. It provides a power for search for stolen and contraband material, and is copied from the Consolidated Customs Act.

Section 78.—Is also new. It is copied from Act XIX of 1873, and provides a penalty for the destruction of boundary-marks.

In conclusion, I have to explain why several of the concluding sections of Mr. Brandis's draft are not included in this proposed Bill. These sections will be found in Part XI of Mr. Brandis's draft (Section 61, &c.)

Section 62—Is included in my section 66.

Section 64—Is purposely omitted. Power to fine subordinate officials is a provision belonging to a service regulation rather than to an Act.

Section 70—I have on consideration omitted. There is no need of a special section. Already the Penal Code makes general provision for cases where public servants vexatiously and needlessly put their authority to exercise in order to annoy, &c.

Section 72—Is needless. The general power of pardon and remission of penalties under the Criminal Procedure Code answers all purposes.

Section 73—Is also needless. There is no object in a special limitation, especially as it is undesirable to add a new provision to the now consolidated statute on the subject.

Section 74.—The same remark applies; and there is no need of the provision. The section *might* cause a gross failure of justice, and there is no corresponding benefit.

Section 75—Is already provided for by the Criminal Procedure Code.

March 31st, 1874.

B. H. BADEN-POWELL,
Officiating Inspector General of Forests.